# The Author

**Liz Bossley** has more than 30 years' of experience in international energy markets, spanning trading, risk management, marketing and extensive involvement of contract negotiations. Liz is the CEO of the Consilience Energy Advisory Group Ltd, which she established in 1999. Consilience includes in its client list major and independent oil companies, utilities, shipping and pipeline transportation entities, regulatory authorities, taxation authorities, trade associations and futures exchanges.

Liz has acted as expert witness in a wide range of more than 25 trading disputes internationally in the field of oil, gas and emissions trading. This has involved appearances in both high courts and before arbitration panels. She is a certified member of the Expert Witness Institute.

She is the principal author of "The Hole in the Barrel", "Trading Natural Gas in the UK", "Bossley's Guide to Energy Conversions", "BFO: The Future Market", "Project Finance Using the Forward Oil Curve", "Climate Change and Emissions Trading: What Every Business Needs to Know" and "Emissions Trading and the City of London". She was the joint author of a report to the G20 in October 2011 on oil price reporting agencies, on behalf of OPEC, the IEA, the IEF and IOSCO. She has been an adviser to both UK HMRC and the Norwegian Norm Price Board on tax reference pricing issues.

**Contributing Editor:** John Walmsley

**Research:** James Walmsley

## Copyright statement

The ISBN Number of this book is 978-0-9550839-4-5

# Chairman's Foreword

In this book the principal author, Liz Bossley, and her colleagues share with the reader many years of experience of crude oil trading. The aim of the book is to provide an understanding of both the theory and practice of buying and selling crude oil and the management and protection of the value of sales realisations using hedging techniques.

The book seeks to serve the needs of new entrants to the trading community as well as the much wider audience of professionals in industry, finance and the regulatory sectors who need an overview of trading operations. They may need this oversight because they work in organisations with trading activities and have a direct or indirect responsibility for them, or simply because they wish to know how the world's largest commodity is bought and sold.

The analysis offered is a realistic and business-like view of oil trading activities, avoiding polemic. While the book aims at utility to the widest possible readership, as Executive Chairman I happily own up to a special fondness for the international independent oil exploration & production sector.

The recent history of technical progress in both exploration and oil field development in the sector is one of almost continual growth and improvement in efficiency. There has indeed also been rapid development of the price management trading tools available to the industry. But it is highly questionable if progress in risk management techniques has been embraced with similar enthusiasm. We do not hear company management reporting as confidently or authoritatively about their new abilities to manage price risk on behalf of owners, as we do about their ability to deploy new exploration and production techniques.

At a conference 10 years or so ago in Houston I had the pleasure of listening to a senior colleague in the industry talk - with numbers - about the successful struggle by an asset team he led to reduce the full cycle costs of a field development by a dollar and fifty cents a barrel. His enthusiasm for the task and the effort that delivery had cost him and his people was manifestly apparent in his face. As a former CEO, I could not help but respect and share his satisfaction with the achievement. Unfortunately, as both a former CEO and CFO with a budgeting background, neither could I prevent myself having two less worthy private thoughts: first, to what extent, if any, would his economies come back to haunt him, or his successor, toward the back end of field life; secondly, what if anything, had he done to assure the actual realisation of his long run sales price

**Tate Library**

assumptions. I had plenty of room for conjecture on this second matter, for although there had been much discussion on unit cost control and technical innovation, there had not been a single comment about price risk management.

It was thinking about this widespread, though to be fair not universal, disparity in the amount of consideration given to the technical and cost issues compared to the time spent on price management issues that led indirectly to the formation of our consultancy, Consilience, and ultimately to the writing of this book. Why should this disparity exist? There is a curious industry fatalism about price realisations stemming from a perfectly correct view that you cannot forecast the long run price of oil. But there is world of difference between frankly admitting, as I do, that I cannot forecast long run oil prices and being a wholly passive price taker.

Examples of current issues discussed include the role of international benchmark, or marker, crude oils in the price formation process. In particular we analyse the position of the Brent marker, made up of a number of North Sea blends, which play an exceptionally significant role in pricing and hedging crude contracts. Some estimates suggest that Brent is used as the marker crude for pricing up to two thirds of the world physical oil sales of approximately 88 million barrels per day. Even more impressively, conservative estimates indicate that Brent is used as the price marker for at least 200 billion barrels of oil per year, both physical and derivative, for hedging and speculation.

Separately there is a discussion of strategic issues for management and Boards concerning price forecasting and investment policy, including the potential role of hedging and its limitations. It is sometimes not wholly understood that the major benefit of hedging is not unit price maximisation, but the provision of certainty and the creation of a secure framework for investment. As discussed later in the book, in this particular context the only "bad" hedge is one that produces a surprise.

The current Dodd Frank and Volcker regulation (soon to be implemented, subject to the provision of detailed regulations that are currently under discussion), welcome in many quarters as a necessary corrective control of derivative activities, has the unfortunate potential for the oil business, particularly the independent sector, of throwing the financing baby out with the bath water. There is no evidence in the oil sector of a general, or structural, misuse of derivatives, or any evidence at this stage that their use has skewed oil prices. But if the anticipated legislation is passed without any exception for the oil industry Dodd Frank will require that many of the current over-the-counter (OTC), i.e. private, unregulated hedging transactions between company and provider, migrate to regulated exchanges and be subject to a regime of margin calls. This will be substantially damaging to corporate cash management.

In a similar fashion the anticipated Volcker rule will limit the ability of organisations to simultaneously make markets, i.e. quote two-way prices, and trade on a proprietary basis. This will initially hit the large investment banks that have provided the hedges necessary to underpin oil field development loans in the independent sector. Sadly, if this comes to pass I suspect the banks will find a faster use for their new spare capacity than the independent oil sector will find new sources of finance - unless new foreign banks move to fill the gap very quickly.

Historically, it has been difficult to attract external funding to both the fixed and working capital requirements of the industry through the various stages from exploration to first oil. This has led over time to the development of a suite of financing techniques developed by and funded within the industry itself, i.e. "industry finance". The overriding objective of the arrangements is to split and spread the project risk amongst the participants best able bear it and to provide suitable rewards for the risks assumed. Examples of such arrangements include farm–outs, carried interest arrangements, overriding royalties and carved out production payments.

For example, where a joint venture group comprising smaller companies has made a discovery, larger companies may be invited into the group to offer balance sheet and technical strength to help the smaller partners participate in a discovery that might otherwise be a risk too far for a small outfits. However, the larger oil companies are acquirers and owners of oil equity, not third party lenders. So such assistance comes at the cost of a larger share in the field and usually with the right of the larger company to trade the smaller producer's oil for it, when it is eventually able to lift.

The development of a banking sector that had oil reservoir engineering expertise and, in some banks, major balance sheet strength, injected external competition into industry finance. The banks also brought to the table a suite of two way tradable instruments that could reduce price risk significantly for the independent oil sector. This has given the independent companies much greater flexibility and the ability in many circumstances to preserve more of their equity in oil field developments. If anticipated financial regulation removes banks from this game it is the independent sector that will be the loser.

For a newcomer to the industry the scale of operation, the complexity of process and the necessary documentation of the physical sale and any insurance, i.e. hedging to protect anticipated cash revenues from the sale, may appear daunting. As ever apparent complexity can be resolved by keeping your eye on the ball.

First, track what is happening to the physical flow of oil; secondly follow the cash; and thirdly, at each stage ask, what would I do if this was my cargo and my cash? From this perspective most of the actions and documentation will fall into place.

Here are the key points to bear in mind:

- Your oil is probably produced into shared storage with joint venture partners in your field and a number of other fields. Everyone has to have a fair opportunity to lift their share and this means that there has to be a schedule that allocates cargoes to companies in time to let each organise a sale.

- That allocation happens typically 4-10 weeks in advance of loading, depending on the particular area of the world in which you are operating.

- A cargo has to be sold at least ten days before it loads to allow the organisation of a ship and to put credit facilities in place with the buyer.

- But regardless of when the cargo is sold the ultimate price you will receive depends on the loading date of the cargo that is allocated to you. Industry convention sells physical cargoes of oil using a formula price in the contract that calculates the price of a cargo by reference to prices published by industry reporting agencies such as Argus or Platts tied to its specific loading date.

- You can either accept the outcome of that price formula and take no further action. Or you can decide to manage it using financial hedging instruments.

If you decide to manage the price you can do so broadly at any time, but a year or more in advance is typical. This management can be achieved by hedging using forwards, futures, swaps or options, far in advance of the date when you know the date of the cargo in question. As the date of delivery gets closer and you know the exact shipping date of the cargo these broad hedges can be fine-tuned using the precision instrument of the CFD, "dated-to-paper" swap.

To assist the reader in getting to grips with these concepts the book uses the Consilience "A/T/G" analysis tool. This unbundles the oil price into three component parts; the absolute price of a benchmark grade (A); a time differential, the price attributed to oil being delivered on different specific loading dates (T); and, a grade differential (G), the difference in price between a benchmark grade and the grade of oil in question.

The construction of the price formula for a physical cargo and the management of the A and T components of that price formula using financial instruments are made clear. The grade differential, G, is explained in detail using assay analysis and by explaining how crude oil performs in the refinery to produce the finished products that the end-user wants.

The early chapters will lead the reader through these issues, but to summarise, the key point for consideration is that the seller or buyer will normally be tied in to a price for a physical cargo related to a price formula, reported by the price reporting agencies on or around the day of loading. However, to help circumvent that restriction and provide price management flexibility, the market offers a choice of tools that allow the hedger to control prices either on a cargo by cargo basis, or on a more strategic level over a longer time frame. Therefore the matter for resolution becomes what are your degrees of freedom to manage your price realisation and how, if at all, do you want to use them?

I hope that you enjoy reading our book and find that it serves its purpose of providing you with a clear and soundly based understanding of how crude oil is bought and sold in practice.

John Walmsley, Chairman

# Table of Contents

## Chapter 1

## Chapter 2

# Chapter 3

# CHAPTER 1

" My ventures are not in one bottom trusted, Nor to one place; nor is my whole estate Upon the fortune of this present year: Therefore my merchandise makes me not sad. "

*Antonio, The Merchant of Venice, Shakespeare*

## Introduction

At periodic intervals throughout my career as an oil trader I have observed with interest each financial trading scandal as it unfolded. From Nick Leeson to Kweku Adibola, from the demise of Lehman Brothers and all that followed to the still unfolding Libor debacle, one theme keeps recurring - when it comes to oversight of the trading function, management is asleep at the wheel.

We have had our own fair share of trading scandals in the oil sector too. Major oil companies still shudder at the mention of Transnor, the small trading company that challenged them in the mid-1980s on the issue of price-deflationary tax spinning and did very well for itself in out-of-court settlements. The Ceylon Petroleum Company's unsuccessful attempt to avoid paying out on 2007/8 loss-making oil hedges to Standard Chartered Bank created more recent ripples throughout the industry.

Senior management of companies may try to claim that it did not know or did not understand what was happening down on the trading desk, but we are finding out that ignorance is a feeble defence at best.

This book is aimed at new crude oil traders or anyone who feels the need for a general understanding of how this crucial business activity works. It is designed to provide a sound framework of understanding of how oil trading functions and how oil price formation takes place in practice.

If you are embarking on a crude oil trading career you will find the vast majority of what you need to know, or where to find it, within these pages. If you are a senior oil industry executive or non-executive, a lawyer, financier, auditor, tax specialist, accountant or project manager and want to understand how oil trading actually works to make sense of the prices at which traders sell or finance oil production, then this is also the book for you.

Trading is a complex discipline; but trust me, it's not as hard as you may fear - as this book hopes to demonstrate. My advice to oil executives would be to do a little bit of work now to get up the trading learning curve. If you are training to be an oil trader you may find it helpful to have a practical guide like this to fill in any blanks that your more experienced colleagues may take for granted and therefore forget to mention to you. Those in other professional disciplines that interact with the traders will find the jargon and basic principles they need to understand the traders and to spot any problems early. If you are responsible for over-seeing an operation that involves trading or if you direct a company where trading takes place as an ancillary activity of the business, reading this book will let you make more informed choices and also help you sleep better.

There is an alternative: you can so constrain the trading function in your company, in the case of an upstream company, that it only undertakes activities and contracts that the board, typically made up of drillers, engineers, refiners, lawyers, and accountants, i.e. those for whom trading is not a first language, understands. That way there can be no derivative surprises that embarrass the company on the front page of the Financial Times or the Wall Street Journal. But that is akin to using stone axes in the age of power tools and it won't be long before your annual report becomes a disappointment to shareholders because money will be left on the table.

It is curious how oil companies will celebrate the success of a 50 cent per barrel annual saving in operating costs while leaving sales revenue to the mercy of an oil price that can, and routinely does, fluctuate by a dollar or more per barrel per day. Make the effort to understand what is happening to the revenue line and you can begin to manage it responsibly and appropriately for the circumstances in which your company operates.

# The Big Picture: the Trading Backdrop

After the 2008/2009 banking collapse the regulators are doing their best to corral trading activity into a measurable and manageable system. The US Dodd-Frank Wall Street Reform and Consumer Protection Act is, amongst other proposed features, feeling its way towards shoe-horning over-the-counter (OTC) opaque derivative transactions onto transparent regulated exchanges where it can be quantified and monitored for systemic risk. All well-meaning and good and it may well work for financial instruments.

But the reason we have an OTC market in the oil sector is because the regulated exchanges only offer a limited range of standardised products that do not match the highly tailored needs of the bona fide oil company hedger. The clearinghouses that manage the exchange user's credit risk do so by calling on initial and daily variation margin payments that are beyond the cash liquidity limits of all but the largest companies. OTC market-makers can currently take a more holistic view of the hedging client's dealing capacity by considering the physical assets being hedged as well as the short-term mark-to-market profit or loss of any hedge positions. Forcing OTC business to be cleared through regulated exchanges will aid transparency, but could put risk management beyond the reach of companies that do not have access to the large, short notice calls for cash that clearinghouses can demand.

The current review of the European Market in Financial Instruments Directive (MiFID) is moving in a broadly similar direction to Dodd-Frank, but is taking its own route and moving at its own pace.

The much maligned Dodd-Frank Volcker rule looks set to stop banks gambling with tax-payers' money, an objective with which we can all agree, by prohibiting so-called "proprietary trading", i.e. dealing on the bank's own account rather than on behalf of clients. Despite what may or may not have been originally intended by this rule, for banks it is crystallising into the maxim "market-making good, risk-taking bad". Unfortunately in the oil sector it is impossible to be a market-maker without taking risks. Consequently, the Volcker Rule could emasculate the banks that offer tailored products to oil companies.

Oil companies that want to take control of the oil prices they will face in the future are obliged to enter into "financial" hedge contracts. That is because it is almost impossible to buy or sell physical oil today at a fixed price if that oil will not be delivered until six months, one year or several years in the future. The market simply doesn't work that way.

The reason hedgers turn to banks to make them a market is because the range of instruments available in the oil sector to lay off complex price risk is inadequate for the needs of the hedger. There are timing and quality differences between the physical oil being hedged and the instruments available to hedge it. If the banks cannot take on these so-called "basis risks" because that would involve them in proprietary trading, those risks may have to remain with the oil companies who are trying to lay them off.

The bottom line is that any oil industry executives hoping that the regulators will make the market safer and simpler for them to use are going to be disappointed. The bigger risk is that the regulators will inadvertently throw out the oil hedging "baby" with the banking malpractice "bathwater". The possible withdrawal of US banks from oil trading will at the very least reduce competition in supplying risk management services to the industry- never a good thing for the cost of sales.

Granted, non-US banks may step into the breach. Many European, Asian, Russian and other banks already operate in this area. In my experience these banks offer hedging services to clients but in many instances can only do so because they can lay off the risk with US banks. Removing the US banks from the picture takes away a key supporting beam from the oil market architecture.

More worryingly, losing the US banks as market-makers in the oil sector is likely to have the consequence of driving oil hedgers into the arms of their major oil company competitors to manage their strategic or project-based oil price risk. Before a market-maker will offer a strategic or project hedge to a company it needs to examine the company's financial position and project data in detail. The OTC market works because it does not currently operate with the clearinghouse system of daily margin calls that cash-constrained companies find so unmanageable. Market-makers operating in the OTC market can offer products to companies by taking into account that company's overall position, i.e. by understanding not only the hedges, but the physical position and contracts that any hedges are designed to cover. Turning project data over to a major oil company, in the absence of derivative market-makers, to support a hedge that may have been imposed on an exploration and production company by the lender of development finance runs the risk of having to sacrifice some equity in the project to the major oil company.

That leaves trading companies with the opportunity to become more active in providing market-making services to hedgers. Increasingly trading companies are moving upstream and into refinery and storage asset ownership, calling into question their objectivity as a hedging counterparty too.

# Simply Selling Oil

Nevertheless the withdrawal of banks from oil trading may not be too worrying – right? After all many small to medium-sized companies have done very well by outsourcing oil trading into long term contracts with the major oil companies in the past. This has allowed them to rest content in the knowledge that their oil will always be lifted safely and regularly at some sort of market average price. No need for all that risky trading and hedging stuff!

In the case of some of the more marketable and actively traded types of crude oil the small oil producers may even be able to pat themselves on the back for achieving premia to the publicly reported market price of their oil by selling to larger oil companies. But there is a saying that if something looks too good to be true, it probably is.

If large oil companies are prepared to pay their smaller cousins a premium over the reported price of a particular type of oil then there is something going on that requires examination, otherwise it may be that there is some money being left on the table by the seller. It could be that there is additional value in existence that is only available to major companies because they have economies of scale and an integrated system that presents opportunities not open to the smaller independents. If that is the case then all well and good and everyone should be happy. Competition amongst buyers may have forced large oil companies to share some of their economic rent with the smaller seller. Enjoy it while the competition lasts. If the regulators force some of the competitors out of the market, particularly US banks, then perhaps those premia will decrease.

But if the premia paid to small independent producers are really a measure of hidden costs to the seller in its profit sharing, cost recovery or tax calculation further upstream, then a bit of effort to understand what is really going on may well pay dividends. We discuss this in Chapters Two and Six.

# The Joy of Trading

Meanwhile down on the trading desk it may be that some readers of this book are simply seeking to understand why they are perpetually being outbid by more sophisticated traders in tenders to buy oil from national oil companies or sell oil to state refineries. Or maybe they are looking for ways to exploit arbitrage opportunities without taking on flat price risk or falling foul of volatile time differentials. This book will unbundle the price equation for you into its component and manageable parts and explain the mechanics of what can be hedged and what cannot.

The book explores the physical crude oil market internationally examining the upstream drivers that explain why oil companies sell their oil production in the intriguing way they do. It assesses the value of different grades of crude oil to refineries and investigates the cascade of agreements that convey oil from the wellhead to the refinery gate. It analyses the anatomy of the oil price considering the benchmark grades in detail, the reason why oil prices vary with their delivery date and why one grade of crude can be worth so much more or less than an ostensibly similar grade.

The book reviews the unique role of Brent and suggests that there may be risks associated with having such a vast volume of international physical oil contracts and risk management instruments built on a very limited number of barrels produced and deals done in the benchmark grade. This is a timely review given that the Brent forward contract is undergoing yet another industry revision just as this book is going to print.

Whether you are reading this book to learn how to be a desk trader or broker, or whether you are reading it to determine what should be the strategic role of trading within your company, or if you need to understand what traders are doing in order to report their activities or even litigate or regulate them, we have tried within these pages to provide a comprehensive, but simple, explanation of what takes place inside the trading silo.

## Conventional Wisdom about the Future

First a bit of context setting. You will not find within these pages the Consilience forecast of future oil prices. There are a number of excellent well-resourced agencies that undertake the thankless task of compiling supply and demand statistics and attempting to predict the unpredictable - the future course of oil prices.

I say it is a thankless task because the closer they come to getting it right, the more likely it is that the unlucky forecasters will turn out to be wrong. This is because the clearer is the future trend in oil prices, the more likely it is that investment money will be deployed in response to the trend, thereby changing the outcome. And that is before you take into account the random geo-political and weather-related shocks that thwart the forecasters' sterling efforts.

Consilience adopts a different approach to the future. What is the point of forecasting at all when there is a forward oil curve that allows those with oil price exposure to act today to construct a future that suits their own plans? We will develop this theme throughout this book. But in the meantime we acknowledge with sincere respect the forecasters' art and will continue to pore over their predictions along with everyone else.

According to the International Energy Agency (IEA) World Energy Outlook 2012, the rate of growth in world Gross Domestic Product (GDP) is assumed to average 3.5% per year over the period 2010 to 2035 in all its scenarios. Non-OECD countries are assumed to have a compound annual average growth rate of 4.8% per annum over this period compared with 2.1% in the OECD region. For China and India these rates are 5.7% and 6.3% respectively.

The scenarios that the IEA outlines can be summarised as follows:

- **The New Policies Scenario** takes into account broad policy commitments and plans that have already been implemented and those that have been announced, even where the specific measures to implement these commitments have yet to be introduced.

- **The Current Policies Scenario** assumes that Government policies that had been enacted or adopted by mid-2012 continue unchanged.

- **The 450 Scenario** is an outcome-driven scenario. It postulates that policies are adopted that would give around a 50% chance of limiting the global increase in average temperature to 2°C in the long term, compared with pre-industrial levels. This would require the long-term concentration of greenhouse gases in the atmosphere to be limited to around 450 parts per million of carbon-dioxide equivalent (ppm $CO_2$ – eq). Hence the title of the 450 scenario.

The outcome of these different scenarios for world primary energy demand is illustrated in Figure 1. The New Policies Scenario is taken as the base case.

## Figure 1 World Primary Energy Demand

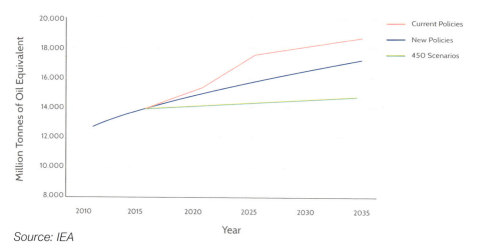

*Source: IEA*

Taking the new policies base case scenario, the difference in the growth rate of world primary energy between OECD and non-OECD countries is quite stark, as shown in Figure 2, with the OECD remaining almost constant. The main growth is coming from China, India and the Middle East according to the IEA's thinking.

## Figure 2 World Primary Energy Demand (New Policies)

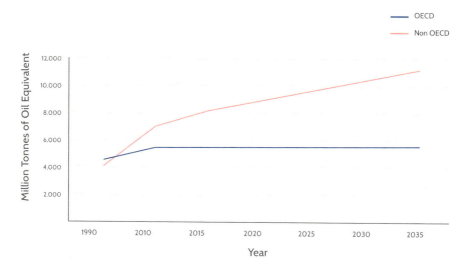

*Source: IEA*

The current base case scenario is that fossil fuels will still form about 75% of primary energy supply by 2035, a drop of about 6% from 2010. It does not take a genius to foresee that the move away from nuclear by Japan, Germany and others could upset this thinking. As the green drive stalls in the wake of an on-going international recession that has left Greenhouse Gas (GHG) reduction permit prices at levels too low to incentivise green investment, there has to be a good chance that the gap left by nuclear will be filled mainly by fossil fuels in the short term and increasingly by renewables in the medium to long term. See Figure 3.

**Figure 3 World Primary Energy Demand by Fuel**

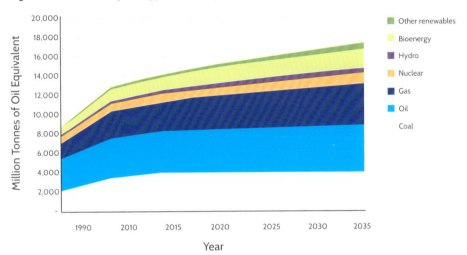

*Source: IEA*

Which fossil fuels are used to meet demand has considerable significance for the green agenda, as shown in Figure 4. Natural gas is the least environmentally unfriendly of the fossil fuels in generating power and heat, according to the Intergovernmental Panel on Climate Change.

**Figure 4 CO$_2$ Emissions From Electricity Production**

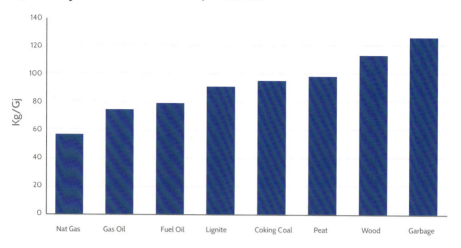

*Source: Intergovernmental Panel on Climate Change*

Nevertheless the high growth plans of China and India in particular guarantee a future role for coal, the apparently least environmentally friendly fossil fuel, as shown in Figure 5, although the coal lobby would challenge the characterisation of coal as a dirty fuel.

**Figure 5 Coal Demand The Big Four (US, China, India, Russia)**

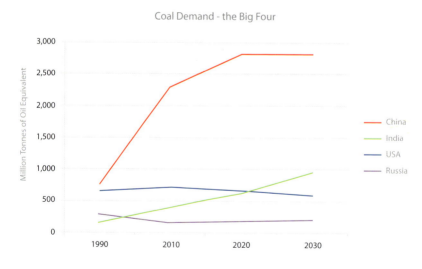

*Source: IEA*

The four countries shown - China, the US, Russia and India - are currently the world's four largest GHG emitters, so when it comes to balancing the needs of energy security with the environmental agenda, the actions of these four are of paramount importance.

**Figure 6 Oil Demand Forecast**

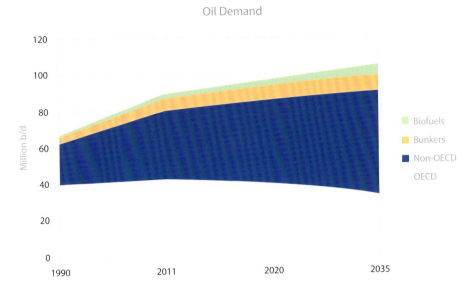

Oil Demand

*Source: IEA*

But this book is about oil and one thing is abundantly clear from the IEA forecasts (see Figure 6) – oil will be a major feature in the energy balance for the foreseeable future. Declines in demand in the developed countries will be more than offset by increases in the developing world, driven mainly by the transport sector.

# Conventional Wisdom about Oil Production and Consumption

We are not going to get into the "Peak Oil" debate in this book. By the time we find out who is right and who is wrong about the point when oil production will pass over the maximum hump into terminal decline most readers of this book will probably be dead. In the meantime the concept does not present many practical trading opportunities. And this book is about oil trading.

Oil is undeniably a finite resource and one that should not be squandered: that is what Americans would call a "motherhood and apple pie" statement. According to that extremely useful publication, the BP Annual Statistical Review, proven oil reserves at the end of 2011 were about 1653 thousand million barrels i.e. 1,653,000,000,000 bbls. See Figure 7.

**Figure 7 Proven Oil Reserves 2011**

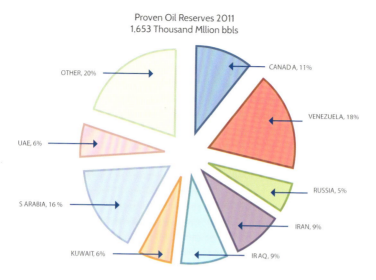

Proven Oil Reserves 2011
1,653 Thousand Mllion bbls

OTHER, 20%

CANADA, 11%

VENEZUELA, 18%

UAE, 6%

RUSSIA, 5%

S ARABIA, 16 %

IRAN, 9%

KUWAIT, 6%

IRAQ, 9%

*Source: BP Statistical Review*

It also reports that 2011 production was 83.6 millionb/d, i.e. about 30.5 billion bbls per year. At current levels therefore proven reserves are equivalent to about 55 years' worth of production. In 55 years not only will new exploration and production techniques have moved many more reserves from the probable and possible to the proven category, but there will also have to have been a surprising collapse in scientific ingenuity if we are still using oil in the same way 50 years from now as we are today. So I am happy to live up to the stereotype of the oil trader as being too "short-termist" by saying that I am not unduly concerned about oil running out.

For the purposes of this trading book I am more interested in where oil is being produced and consumed and the implications for transportation from areas of supply to areas of demand. The BP Statistical Review is again helpful with its basic data. (See Figure 8) Saudi Arabia and Russia have been vying for the title of the world's largest oil producer for the last few years, with the former regaining the title in 2011. Both the US and China were each producing about 41-42% of their own needs domestically in 2011, but as Figure 6 suggested, non-OECD demand growth led by China is likely to be a dominant feature of the trade pattern in the coming years. We should also not ignore the American effort to boost energy self-sufficiency, which is leading some pundits to suggest that the prospect of US crude oil exports could soon be taxing US politicians.

**Figure 8 Oil Production and Consumption 2011[1]**

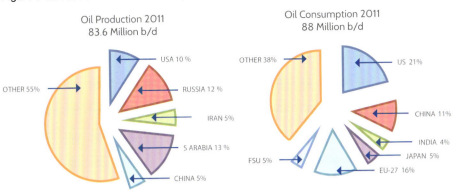

Oil Production 2011
83.6 Million b/d

USA 10 %
RUSSIA 12 %
IRAN 5%
S ARABIA 13 %
CHINA 5%
OTHER 55%

Oil Consumption 2011
88 Million b/d

OTHER 38%
US 21%
CHINA 11%
INDIA 4%
JAPAN 5%
EU-27 16%
FSU 5%

*Source: BP Statistical Review*

Diversification of supply sources is at an advanced stage in both of the two largest importing regions, Europe and the US, and also in China. Supply is remarkably undiversified in the case of Japan (See Figure 9).

While Europe is still dependent on Former Soviet Union countries for about half of its crude oil imports, the FSU is itself now a diversified region with countries such as Azerbaijan, Kazakhstan and Turkmenistan producing moderately significant quantities of oil in their own right. Nevertheless the interconnectivity of the pipeline system that requires the transit of several borders including Russia, Georgia and Turkey and in some cases the Bosporus bottleneck, to get FSU crude to market makes it difficult to consider any of these countries in isolation from its neighbours.

The US imported more than 50% of its needs from its immediate neighbours in the Americas in 2011. Increasing quantities from Canada delivered by pipeline straight into the US ticks many political and economic boxes for both participants in that trade. The US is less likely to raise awkward questions about the environmental credentials of Alberta tar sand oil that has been instrumental in causing Canada to fail to meet its Kyoto GHG reduction target. For the US a secure supply from a like-minded neighbour is undoubtedly very welcome.

---

[1] Consumption exceeds production partly because of stock changes but mainly because the former includes inland demand plus international aviation and marine bunkers and refinery fuel and loss. Consumption of fuel ethanol and biodiesel is also included.

## Figure 9 Diversification of Supply in the Major Importing Regions

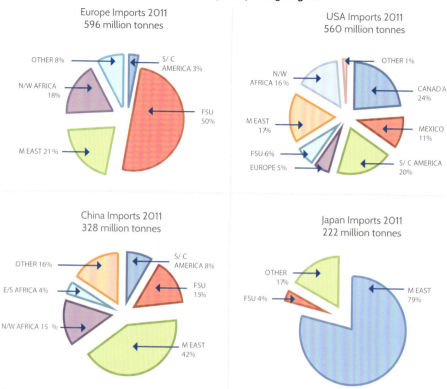

Source: *BP Statistical Review*

The US runs the risk of over-playing its hand in that relationship. Deliveries of Canadian crude have suffered low prices over the last few years as US domestic prices have struggled with a logistical bottleneck at Cushing, Oklahoma, the main delivery point for the Chicago Mercantile Exchange's NYMEX light, sweet crude oil contract. From trading historically at a $1-2/bbl premium to its rival seagoing marker grade, Brent Blend, the price of this US benchmark grade has traded as low as $25/bbl under Brent, delivering disappointing results to Canadian producers. Over time investment in infrastructure should solve this problem - already the reversal and extension of the Seaway pipeline from Cushing to the Gulf Coast is alleviating some of the pressure, reducing the discount to circa $15-18/bbl.

If the extension of the Keystone pipeline from Cushing to the Gulf Coast goes ahead, Canadian crude will be able to access the international market from Alberta, via Montana, South Dakota, Nebraska, Kansas and Oklahoma and out through the Texas coast. But the pipeline extension is proving controversial and early in 2012 President Obama postponed its construction indefinitely.

Meanwhile back in Canada the Enbridge Northern Gateway pipeline project to export Alberta tar sand oil across British Columbia to the Pacific, where it would doubtless be warmly received by the Japanese, South Koreans and the ever-hungry Chinese, has yet to receive environmental approval.

It is difficult to make any sensible comment about Chinese trade patterns because they are evolving so quickly. Future imports are as likely to be determined by the country's acquisition of oil producing companies, for example Talisman and Nexen, as it is by it negotiation of purchase contracts.

The construction of the Russian East Siberian Pacific Ocean (ESPO) pipeline with its spur direct into Daqing delivers 300,000b/d direct to the Chinese refinery. This appears to be a sensible strategic move, but the relationship has been rumoured to be fraught with tension.

Alas for unlucky Japan. Having reduced its oil fired electricity generation from 72% in 1973, when the first oil shock sent its economy reeling, to 7% in 2009[2] the tsunami and ensuing disaster at the Fukushima nuclear facilities in March 2011 has sent it right back to square one. For a country with next to no natural resources, tough Kyoto commitments that it has honoured to the hilt when others have failed, and which was already making extensive use of renewable energy sources, the blow is a bitter one. It will take some time for a new energy policy and stabilised trade patterns to emerge.

## OPEC and National Oil Companies

It would be wrong to embark on a book about oil trading without mentioning the Organisation of Petroleum Exporting Countries (OPEC). Despite the fact that OPEC meetings do not make the 6 o'clock news the way they did in the 1970s and 1980s, OPEC remains a force to be reckoned with. This is likely to be increasingly the case as the fortunes, and spare capacity, of Iraq are restored in the coming years.

Apart from the mid-1980s, when OPEC's share of world oil production fell to a low of 27% in 1985, sparking off the netback pricing war of 1986, OPEC's share of world production has stayed remarkably constant at around 40%, give or take a few percentage points (See Figure 10). No matter how diversified is the import portfolio of a developed country the policy decisions of producers controlling 40% of the world's oil will have an impact on the size of its import bill.

---

[2] The Journal of Energy Security Dr. Vlado Vivoda December 2011

For many, OPEC is synonymous with the Middle East, but in fact the current line up of members is Algeria, Angola, Ecuador (who left in 1992 then re-joined in 2007), Iran, Iraq, Kuwait, Libya, Nigeria, Qatar, Saudi Arabia, the United Arab Emirates and Venezuela. Indonesia was a member, but it left in January 2009 after it became a net importer. Gabon was a member until 1995.

The stated mission of OPEC is "to coordinate and unify the petroleum policies of its Member Countries and ensure the stabilization of oil markets in order to secure an efficient, economic and regular supply of petroleum to consumers, a steady income to producers and a fair return on capital for those investing in the petroleum industry".

OPEC attempts to influence the international price of oil by allocating production quotas to its members, but has had mixed success in agreeing these quotas and in enforcing them.

**Figure 10 OPEC Share of World Oil Production**

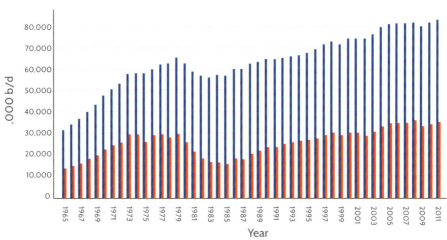

*Source: BP Statistical Review*

As a starting point for analyzing oil prices from a trading perspective the OPEC basket price (See Figure 11) is only of passing interest as a broad measure of international price levels. The basket was introduced in June 2005 and is currently made up of Saharan Blend (Algeria), Girassol (Angola), Oriente (Ecuador), Iran Heavy (Islamic Republic of Iran), Basra Light (Iraq), Kuwait Export (Kuwait), Es Sider (Libya), Bonny Light (Nigeria), Qatar Marine (Qatar), Arab Light (Saudi Arabia), Murban (UAE) and Merey (Venezuela).

**Figure 11 OPEC Basket Price**

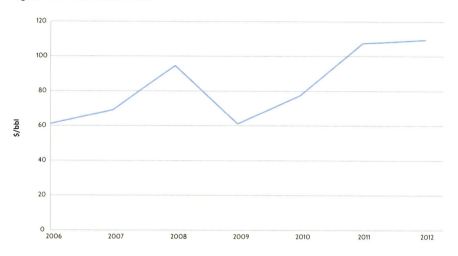

*Source: OPEC*

Consideration of this basket at last brings us to the question of oil prices.

## The Evolution of Pricing Conventions

In the 1970s and early 1980s the price of oil was typically expressed as a "fixed and flat" price, such as $X/bbl[3]. In those days prices were negotiated and agreed usually on a quarterly basis between predominantly major and large independent oil companies on one side and OPEC and other large oil producing countries on the other[4]. There was no real concept of the forward or futures market showing the price today at which oil could be bought or sold for delivery in future time periods. If there was a forward oil price curve at all it was a straight horizontal line stretching up to three months into the future.

There are many learned texts and theories as to why this rigid price structure broke down. My perspective on the change is derived from my experience as an oil trader for the British National Oil Corporation (BNOC), the UK state oil company, at that time. BNOC represented one of the non-OPEC producing countries that competed with OPEC to sell its domestically-produced oil to the industry at market-clearing prices. Unlike the OPEC producers, countries like the UK and Norway did not attempt to influence

---

[3] For a look at the price formation process further back in history see "An Anatomy of the Crude Oil Pricing System" by Bassam Fattouh, January 2011

[4] A notable exception was and still is the US domestic oil industry which operates using prices "posted" by buyers or aggregators setting the price and other terms on which they will buy oil from independent producers.

prices by adjusting volumes and had to sell all the oil it produced, or more correctly was produced for it by the oil industry under license agreements, called Participation Agreements, each quarter.

BNOC and its Norwegian counterpart, Statoil, therefore had to arrive at a quarterly negotiated price at which it was prepared to buy oil from the industry under its participation agreements. These state entities had to be prepared to resell this oil, together with its own equity share of production and royalty oil taken in kind, back to the industry at the same price as the price at which it purchased it from the licensees. Initially these sales were made under term contracts of typically one year in duration. In the early days of its existence BNOC had no government approval to sell oil into short term spot contracts to balance supply and demand. Therefore the level at which it set its quarterly price had to be finely tuned to exactly match demand with varying supply.

As non-OPEC supply took an increasing share of the world market in the first half of the 1980s (See Figure 10) OPEC brought increasing political pressure to bear on non-OPEC national oil companies to support its bid to raise the price of oil. UK treasury buckled under this pressure and induced BNOC to fix its quarterly prices at levels greater than those at which it could re-sell all its oil under annual term contracts. The only way this could be accommodated was by selling surplus cargoes into the spot market, usually to trading companies. Before this time the spot market was commonly quoted to be only about 10% of international production, although the data to verify this statistic was and still is hard to come by.

There was considerable prejudice against selling to trading companies, who were often dismissed as parasites who extracted an ill-deserved margin from the coffers of "real" oil producers and consumers. Traders were given scant credit for the vital role they played in managing price risk and balancing supply and demand. They filled the role of industry whipping boy that is today reserved for speculators, a term which is defined by the Oxford English Dictionary as one who "invests in stocks, property, or other ventures in the hope of gain but with the risk of loss", but is often used more pejoratively.

OPEC producers also sold increasing quantities to trading companies in the spot market, the price of which began to crystallize in industry publications such as Argus and Platts who reported prices in the spot market daily. These spot prices were considerably less rigid than OPEC posted prices and non-OPEC negotiated prices and thus oil price volatility became a feature of the market.

In 1981 the first forward oil contracts at fixed prices began to be traded and the forward oil price curve was born. This allowed both producers and refiners not only to balance supply and demand in advance, but to manage their price uncertainty. It provided one of the first risk management tools where traders could lay off some of the price risk they had taken on from producers and refiners. This supported the further growth of the spot market.

Buyers in long term contracts at high fixed prices were reluctant to renew them as they identified increasing quantities of oil trading in the spot market at lower prices. But many were unwilling to abandon the security of having an oil supply guaranteed by a national oil company. The solution was obvious: long term contracts for physical supply with the prices "floating" and set by reference to published spot prices relevant to the delivery month of the cargo in question. By 1985 the era of the formula price had begun and the scene was set for increasing complexity and decreasing transparency in the price formation process.

The next layer of complexity was the emergence of formula pricing for spot cargoes, the price of which became related to the fixed price of the forward contracts as published on or around the loading date of the spot cargo.

Today the majority of physical oil cargoes are priced by reference to formulae, which vary in detail between long term contracts and one-off spot contracts:

- In long-term contracts, which usually mean contracts of a year or longer in duration, the price of each cargo delivered under the contract will often be expressed as the average of published price quotations over the calendar month of delivery of each cargo.

- In spot contracts, i.e. contracts for the delivery of usually one identified cargo in a specified date range, the price is often expressed as the average of published quotations on the 3-5 days around the anticipated loading, or bill of lading (B/L), date. In some regions, for example West Africa and parts of Eastern Europe, the spot price is expressed as the average of published quotations on the 3-5 days after the anticipated B/L date.

A few more recent term contracts are beginning to adopt the spot contract pricing convention of pricing the oil by reference to the B/L date of each cargo delivered rather than as a monthly average.

# Important Pricing Convention

By convention, and only by convention, the trading community considers such a spot formula to deliver the objectively "correct" value of the cargo in question. The practice of pricing by reference to the B/L date is wide-spread, but it is still evolving and regional variations are becoming more apparent. These regional variations often reflect how the national oil company or other state entity responsible for administering the upstream production license framework or taxation regime chooses to define "the market price". We will revisit this production license and taxation framework in Chapter Two and consider the implications for trading practices.

## The Three Components of the Oil Price Formula

The analysis contained in this book will use terminology that reflects the fact that the price of any given cargo of crude oil is typically made up of three different components. These are:

- the Absolute Price (A);

- a Time Differential (T); and,

- a Grade Differential (G).

This is represented in Figure 12, which depicts the forward oil price curve. The component, A, the absolute price, is the height of the forward price curve for a benchmark grade of oil; the time differential, T, is the position along the forward price curve defined by the delivery date of the oil in question; and, the grade differential, G, is the distance off the forward price curve representing the difference in value between the benchmark grade of oil and the grade of oil in question.

This forward oil price curve is not a forecast of what prices will be at some future date. Instead it is a snapshot taken at a particular instant in time of the prices at which buyers and sellers are actually prepared to deal at that moment in oil for delivery at different dates in the future.

In Figure 12 we show a market in "backwardation", i.e. the curve slopes down from left to right meaning that the price of oil for delivery tomorrow is more expensive than oil for delivery next week, next month or next year.

**Figure 12 Forward Oil Curve in Backwardation**

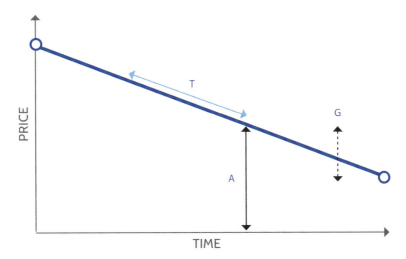

In Figure 13 we show a market in "contango", i.e. the curve slopes up from left to right meaning that the price of oil for delivery tomorrow is less expensive than oil for delivery next week, next month or next year.

**Figure 13 Forward Oil Curve in Contango**

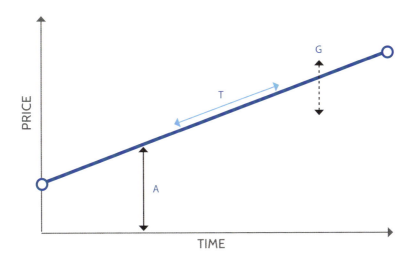

# Consilience Terminology A, T and G.

A/T/G is Consilience terminology that we have been using for a number of years and in a number of publications. However, if you ask a trading company what values of A, T and G it is using in constructing a price formula you will be greeted by puzzled looks. Although traders do not use this terminology they would instantly recognize the concepts they represent. Ask any trading company what is the price of a cargo and it will talk in terms of a benchmark grade (A), the delivery window of the benchmark relative to the delivery window of the cargo in question (T) and the price difference between the benchmark grade of oil and the grade of the cargo in question (G).

But before we start to discuss the anatomy of prices in any detail it is first necessary to understand the contractual process whereby oil is produced, transported and allocated to oil producing companies for sale to the market and ultimately to refineries before we can discuss physical oil contracts and the price formulae they contain. That will be the job of the next Chapter of this book.

# CHAPTER 2

*❝ A verbal contract isn't worth the paper it's written on ❞*

*Samuel Goldwyn*

## Introduction

Like comedy, the secret of oil trading is all in the timing. There are certain practical timetables that influence how and when cargoes of physical oil are offered for sale in the market. These timetables are connected with the upstream forecasting of crude oil field production, the allocation of produced oil between governments and private sector exploration and production companies, the transportation of oil via pipelines into shared storage facilities at onshore or offshore terminals for onward delivery out into the wider market in time to meet refinery purchasing schedules. Similarly the practical considerations of chartering oil tankers to transport oil from production facilities to refineries, the documenting of those cargoes and the arrangement of credit facilities, which guarantee payment, all influence the deadline by which cargoes must be sold. These timetables determine the shape of the sale and purchase agreements for cargoes of crude oil.

This Chapter will look at the cascade of agreements from upstream licence agreements all the way down to tanker charterparties to explain why physical oil cargo sale and purchase contracts have evolved into their current form. We will also describe what the sale and purchase contract for a physical cargo of crude oil looks like.

## Licence Agreements and Production Sharing Contracts

Exploring for oil is a costly and risky business and a highly specialised skill. Consequently the countries that wish to establish if they have oil (and/or gas) reserves within their national borders or under their territorial waters frequently outsource this risk to the private sector. Many do this by licencing tranches of acreage to major and independent oil companies[5]. Exploration contracts form a fascinating area of discussion in their own right. But for the purposes of this trading book we will restrict ourselves to highlighting those issues that have an impact on the eventual negotiation of cargo sale and purchase agreements.

These licence or concession agreements vary from country to country, but are usually granted by a government, or government agency. They typically commit the licensee, or contractor, to shoot a minimum number of seismic surveys and/or drill a certain number of exploration wells within a specified period of time[6]. If they do not do so there are typically break clauses that oblige the contractor to relinquish the licence if they have either not fulfilled their commitments or if they have carried out the work, but failed to find hydrocarbons. In some cases the contractor may have found hydrocarbons, but in quantities it considers uneconomic to develop. In these cases the licence may have to be relinquished to a different contractor with a more economic plan for developing the reserves or who has a lower expectation of returns.

As with most contracts competition will decide the "price" of these licenses. Acreage that is highly prospective, for example adjacent to an area where oil has already been found or showing a geological configuration similar to one where oil has been found in another country, will be much sought after. Hopeful explorers may be required to pay substantial signature bonuses to the government to obtain the licence to explore choice acreage. These licences are often allocated in licencing rounds, in some cases with awards being made to the company that commits to carry out the most work in the shortest period of time. Or the best acreage may be awarded to a company that agrees to also take on a commitment to explore acreage where the likelihood of finding oil is pretty low, as a condition of the deal to gain access to the acreage the contractor really wants.

In countries where exploration risk is minimal and domestic expertise is well-developed, for example certain Middle East countries, contractors may be offered simple service contracts. In such cases the contractor renders services and is remunerated usually in cash. In other cases the contractor may be allowed to own the oil reserves it locates on

---

[5] See " Contracting and regulatory issues in the oil and gas and metallic minerals industries" Mike Likosky, April 2009

[6] As with many issues in this book the US proves an exception to the general rule. In the US private landowners have the ownership right over any minerals found beneath their property, although surface rights and mineral rights may be unbundled and sold separately.

behalf of the host country.

It would be very unusual for an oil company to commit to incur exploration expenditure without knowing in advance what will be the host government's terms for allowing the oil company to develop and sell any oil reserves it actually finds. It would be very poor negotiating tactics to wait until oil has been discovered to agree how much the oil company will be allowed to keep and for how long it would be allowed to retain its production rights. So it is typically the case that the concession regime will be framed within the context of a law that includes a standard national form of Production Sharing Contract (PSC), or Production Sharing Agreement (PSA).

The basic concept of the PSC is that contractors provide the cash to explore and develop licensed areas. Once an oil field is on-stream, the host government takes a share of the production either in cash or in kind[7], or in some combination of both. The host government's interests are usually represented by a national oil company (NOC), who, in some cases, may also have an equity interest, i.e. a percentage share, in the licensed acreage.

## Cost Recovery

Exploration, development and production costs incurred by contractors under PSCs are usually recoverable from the production when it comes on-stream. In other words the contractor can keep money back from the sales of oil produced to recoup its previous expenditure. In most cases the amount of costs that can be recovered are capped at a proportion of the revenue that arises in the first few years: costs not recovered in year one are recovered in years two or three etc. This cap may be anything from 25%-75%, but rarely outside this range. What qualifies as a legitimate recoverable cost is usually spelt out in considerable detail in the PSC and companies are put to proof of actual expenditure.

## Profit Share

Once costs are recovered, profits from the project are split between the contractor and the host government in accordance with the percentages also set down in the PSC. This percentage is again very variable within a range of about 25-75%. In some cases the split is implemented in kind, rather than in cash, by each party having the right to "take and separately dispose of its profit oil share". Usually the PSC will contain an agreement to agree a lifting procedure that allocates cargoes of oil between the

---

[7] In the form of a share of oil production.

contractor and the state oil company. In some cases the contractor is required to sell the host government's profit oil for it, usually by allocating a fixed percentage of the revenue from each cargo it sells to the government's account.

In some regimes, the host government's profit share may be handled by giving it an equity stake in the project to the NOC. This is akin to the practice of giving a "carried interest" in a project to a partner who does not participate at the exploration and development stage, but who is entitled to a share of production[8] once the project is on-stream.

## Royalty and Tax

Royalty is effectively a tax on production, which can be taken in cash or in kind, before or after cost recovery. Typically this is less than 20%. Tax can be levied either on profits or on sales revenue. This is, in most cases, taken in cash, not in kind. Even the broadest of indications of percentage tax rates would be misleading as they can be extremely variable.

## Domestic Obligations

In addition contractors may be obliged to accept other obligations under a PSC. These may be:

- **Supply.** Although the principle that contractors have the right to take and separately dispose of oil in the export market is firmly entrenched in most PSCs, there may also be a provision that the contractor has the obligation to supply the domestic market with crude oil, or, on some rare occasions, with refined products, if the host country does not have domestic refineries. This is typically an obligation to sell oil at "market" prices to domestic companies.

- **Other.** There may also be a requirement to use domestic contractors and materials and to employ local nationals for work undertaken: this may be considered a hidden cost, although it should be recoverable under the cost recovery procedures. The obligation to develop domestic capacity, i.e. to train local oil company executives, is an increasing feature of the oil industry. There may also be obligations to build roads, schools or other infrastructure in the host country.

---

[8] The practice is sometimes used by oil companies who have a licence agreement, but do not have the funds to carry out the exploration and development work.

For readers who are natural traders the opportunities presented by the PSC model described to finesse the contractor's realisations are obvious. So obvious in fact that most PSCs have provisions to protect the host government's position against such tactics. The first of these is the concept of the "ring fence".

## Ring Fencing

PSCs may govern more than one licence in the country in question in which the contractor is involved. When considering the profitability of an individual oil field, there may be an effective "ring-fence" applied to ensure that oil sales from a field are used to recover only those costs that are directly attributable to that field. If this were not the case the allocation of exploration costs from a different block or licence to a project in production could delay "payback" indefinitely. Until payback, i.e. the total recovery of all allowable costs, is achieved, the government's profit oil share is reduced. In some regimes, to further incentivise exploration, the host government may allow such other costs to be applied to a producing project, which has the effect of reducing its profitability. Contractors are required to account meticulously for such "imported" costs.

## Market Price

Host governments and NOCs are increasingly aware of the techniques, including the hedging techniques described later in this book, which could be used to depress the market price realisations that are reported to them by contractors. If the price at which the contractor sells its oil is low then it will take it longer and more barrels to recover its costs. This will in turn reduce the host government's profit share in the barrels produced and will lower royalty and tax payments. If the contractor is selling the host government's share of the oil for it, government revenue will again fall short.

Early PSCs were initially alerted to this issue in the context of non-arm's length sales, i.e. sales where the contractor had an interest in the proceeds of the cargo. Non-arm's length sales may arise because the oil is being sold to the contractor's own refining affiliate. Or a sale may be non-arm's length because the oil is being swapped with a third party for a different grade of oil. In such cases the contractor has less concern in ensuring that the absolute price of the cargo being sold is fair than in the price differential between its sale and purchase under the cargo exchange arrangement. Typically the PSC market price is for an FOB sale, i.e. there is no shipping involved in the price to muddy the water with suspicions that the FOB element of the price is being depressed and the freight element inflated to cheat the host government.

Most modern PSCs tend to have a reference to an oil price that is in some way related to prices obtained in the international market at arm's length between a willing buyer and seller, rather than in the price actually achieved by the contractor itself for its own sales contracts.

Nevertheless ensuring that the market price that is used to calculate cost recovery, profit share, royalty and tax is fair to both parties is an on-going challenge. The challenge is greatest when the oil in question is a type that is not actively traded in the market. This may be the case when the contractor is a large integrated oil company that takes the whole or a large part of the production stream into its own refining system, so that there is little or no independent market price evidence on which to base the PSC market price. In those cases it is difficult to assess what a truly arm's length price would be if it existed.

The issue of the market price clause in PSCs takes us straight into the heartland of the topic of this book. When encountering a new grade of crude oil or when buying oil from a different PSC regime for the first time the prudent trader would be well-advised to study the market price clause in the PSC. This often dictates how the crude is traded. If the PSC market price is based on, say, monthly average published price data, any contractor that sells its oil on a different basis, say based on 3-5 day average published price data in accordance with standard trading practice, will be taking on a substantial risk. For example, Table 1 shows the monthly average crude oil price in 2012.

## Table 1 Price Risk in Production Sharing Contracts

| $/bbl | Monthly Average Price | Maximum 3 day average | Minimum 3 day average | Range |
|-------|----------------------|----------------------|----------------------|-------|
| Jan-12 | 110.50 | 1.75 | -2.25 | 4.00 |
| Feb-12 | 119.50 | 8.25 | -5.25 | 13.50 |
| Mar-12 | 125.25 | 2.00 | -2.25 | 4.25 |
| Apr-12 | 119.50 | 3.00 | -5.00 | 8.00 |
| May-12 | 110.00 | 9.00 | -8.25 | 17.25 |
| Jun-12 | 94.75 | 5.75 | -5.50 | 11.25 |
| Jul-12 | 102.50 | 5.75 | -4.25 | 10.00 |
| Aug-12 | 113.25 | 6.75 | -3.00 | 9.75 |
| Sep-12 | 113.00 | 3.50 | -3.75 | 7.25 |
| Oct-12 | 111.50 | 4.25 | -4.00 | 8.25 |
| Nov-12 | 109.00 | 1.75 | -1.75 | 3.50 |
| Dec-12 | 109.50 | 1.75 | -1.25 | 3.00 |

*Source; Consilience*

A PSC market price definition might well use the monthly average number shown in the first column of Table 1 to calculate cost recovery, profit oil and tax. If however the contractor is using the common trading practice of selling its oil based on a 3-day average price, it might find that its liabilities under the PSC have been calculated on an entirely different revenue stream than the one the contractor actually receives.

To illustrate this risk Table 1 also shows for each month of 2012 the maximum and minimum prices that the contractor might actually have received in each of those months from selling on a 3-day average price basis. So, for example, in May 2012 the PSC calculations may have been based on a price that could have been $9/bbl above or $8/bbl below the price at which the contractor sold its oil in that month - a huge windfall profit or unexpected loss for the contractor.

It may be considered that the contractor ought not to worry about such fluctuations because over time they will probably cancel each other out leaving the contractor broadly neutral in the long run. In some regimes that may well be true. However, in the short term the contractor may receive some nasty surprises for which its cash planning may be unprepared.

Furthermore, if the scheduling of the entitlement to lift cargoes is in the hands of a company, which may even be the NOC, that cherry picks the best dates for itself each month, the contractor may find that it is more often on the wrong side of such price variations than a normal distribution curve of events might lead it to expect. This is because the price of a cargo of oil varies with its expected delivery date, as we will explain in some detail in Chapter Four.

If the operator of the lifting terminal is one appointed out of a group of joint venture (JV) partners, the terms of the Joint Operating Agreement (JOA) should outlaw any such cherry-picking in principle. However what actually happens in practice may warrant closer examination.

## Joint Operating Agreements

The "Contractor" under a PSC on many occasions is not a single company, but a group of companies each with an equity ownership stake in the project in question. Typically interface with the host government, negotiations of service contracts and the administration of the PSC is conducted on behalf of JV partners by an "operator" appointed under the terms of the JOA. The operator carries out the wishes of the JV group in such matters as organising the drilling of wells, designing and gaining consent for a field development plan, negotiating transportation agreements and terminal lifting agreements and all on-going field maintenance and operations. The operating plan and budget is discussed and approved by all the JV partners in regular JV operating committee meetings. Before committing to expenditure over an agreed threshold amount, the operator is required to gain formal authority for expenditure (AFE) from the JV partners.

The operator is usually the JV partner with the biggest equity stake in the field because the appointment is subject to a vote. In some cases the operator might instead be a minority partner who has particular experience of the specific technology needed or might have a long-standing good relationship with the host government of the country in question. The JV partners have the ability to vote the operating company into its position and have the right to vote it out again, usually requiring a majority towards the upper side of a range of between 51% and 75% of the ownership interest, if it is not conducting the group's affairs efficiently *and impartially.*

One of the requirements of a "good and prudent operator" is that it acts for the benefit of all its JV partners. The actual personnel involved in carrying out operations for the field often sit behind a "Chinese wall" that requires it to treat its own colleagues in its affiliated or parent company as third parties showing them no special privileges compared to other JV partners.

This principle extends to the allocation of the right to lift cargoes, which is why the JOA requires mention in this oil trading book. In Chapter One we introduced the concept of a time differential (T) component of the oil price. This illustrates that the value of a cargo of oil varies substantially depending on its delivery date. The allocation of cargo lifting dates therefore has significant economic implications for the JV partners and for the NOC that acts on behalf of the host government.

The JV partners may oblige the operator to market crude oil produced from the field on behalf of the whole group. JV partners that are small companies without in-house trading capability may prefer to make the marketing of JV oil a responsibility of the operator. The smaller companies consider that they are protected by the JOA Chinese Wall that separates the operator's JV personnel from those of the rest of the operator's activities. This aims to ensure that the operator acts as impartially in the allocation of cargoes and sales transactions as it does in awarding contracts for the construction of the oil producing platform. In theory, any operator that favoured its own parent company in the allocation of oil sales would risk being voted out of the position of JOA operator.

However if the operating company has a majority stake in the oil field it can be tricky in practice to vote them out of office, particularly if the smaller companies do not wish to lose the operator's expertise in other areas for the sake of possible revenue foregone, i.e. opportunity cost, in the oil marketing activity. But depending on the PSC regime and the exact nature of the operator's terms for selling oil on behalf of the JV, these opportunity costs may in fact be actual costs of the nature and size described in Table 1 above.

Small companies without in-house trading expertise are almost by definition incapable of identifying and proving that the operator is favouring its own parent company in the allocation of cargoes. Similarly the operator's JV personnel sitting on one side of the Chinese wall may not be aware of any such bias, usually because the actual nomination of dates to lift cargoes and the selling of those cargoes is carried out by the parent company's oil traders sitting on the other side of the Chinese wall.

What form any such bias might take will become more apparent as we develop the themes in this book, particularly as we consider in detail the nature and size and value of the time differential (T) in the price equation.

The most overt form of bias, and consequently the least likely to succeed, is for the operator, if it is vertically integrated, to sell cargoes of JV oil into its own refining system at prices that under-perform what a truly arm's length price would be. The JV partnership may in all likelihood prefer the operator to run tenders amongst a wide range of potential buyers to demonstrate that the price achieved is the best available for all the JV. However, with a majority stake in an oil field, the operator would be entitled to expect at least the same rights as the other partners, i.e. to run its own oil in its own refining system if it so desires, regardless of what other buyers in the market might be prepared to pay.

A more subtle form of exposure for the minority partners, and one which may not even qualify as bias, is when the company selling the oil on behalf of the JV sells it at the best price available, but one which opens up basis risk between the sales price achieved and the market price under the PSC. The operator's trading colleagues can hedge its own company's basis risk, but may not choose, or more likely will not be permitted under the terms of the JOA, to hedge the other JV partners' basis risk.

Few JV partnerships would support the cost of hiring separate traders solely dedicated to marketing JV oil and, even if they did, would probably shy away from giving any such traders authority to hedge. Small companies who want the operator to take physical oil cargo trading responsibility away from them are probably the types of companies for whom any form of trading in forward, futures or derivative instruments is unwelcome. However they may be less concerned about paying the operator's trading affiliate a marketing fee of up to several tens of cents per barrel to sell oil on their behalf. Marketing fees of a dollar or more are not unheard of.

The host government NOC may be indifferent to who markets the oil and to the sales prices the designated marketer achieves, because the NOC has its own protection in the PSC. JV obligations to the NOC in the PSC may be calculated and discharged at the market price as defined in the PSC, not at the price actually achieved by the operator in selling the oil. The NOC often has no obligation to pay any marketing fee that is levied under the JOA, because in a lot of cases the NOC is not a signatory of the JOA. Its rights and obligations are typically set down exclusively in the PSC, not the JOA.

How big any such marketing fee is will depend on many factors, such as:

- whether the parent of the operator actually buys the oil from the JV partners or simply sells on its behalf; or,

- when and where risk and title to the oil passes from the JV partners, either to the operator's parent company or to third party buyers; or,

- whether the oil in question has particularly difficult handling characteristics, such as high acid content or a high pour point or metal contaminants; or,

- whether minority partners would have only infrequent cargoes if they accumulated, lifted and sold their own cargoes. Partners with a small interest that allows them to lift only a couple of times per year may wish to even out their cash-flow and spread their price risk. These will tend to favour selling to or through the operator in order to have some barrels on each cargo that is sold out of the field; or,

- whether the oil field logistics (e.g. daily production rates relative to available storage capacity, or any other constraints that limit a wide range of tankers from loading at the oil terminal) make the job of selling the oil an especially onerous and time-consuming one.

But the over-riding factor that determines the level of market fee is the capability of the other JV partners to take their own oil and sell it themselves if the operator is, in their opinion, charging too much. If there are two or more big companies with an equity stake in a field, it is unlikely that the operator will be required or allowed to market on behalf of all the JV partners. In this situation any minority companies involved in the JV enjoy the more comfortable position of having the larger companies compete to either buy the minority company's oil or market it on its behalf.

## Transportation Agreements

The price achieved for oil in the market is heavily dependent on how, where, when and in what quantities such oil is made available for sale.

Once oil has been discovered and the decision has been made to develop the field, getting the oil to market as efficiently and cheaply as possible is a key component of the development plan. It is at this point that exploration and production companies can make serious errors in concentrating solely on getting the oil on-stream with the minimum of upfront capital expenditure (CAPEX). CAPEX hits the cash flow immediately in an obvious and easily quantifiable form in terms of a reduction in net present value

(NPV) of the field economics, so it is unsurprising that this is the main focus of attention.

What is often ignored or under-valued is the on-going impact by way of increased operating costs (OPEX) on the future revenue stream of a development plan that does not take full account of its trading and price implications. The major oil companies are well aware of this and would not approve a low-cost development plan that did not value accurately any loss of trading flexibility and the company's future ability to optimise price.

## Tariffs and Conflict of Interest

The choice of transportation route can be fraught with issues of conflict of interest. Conflicts of interest that are easily identifiable may be dealt with under the terms of the JOA. For example, if a field is in the happy position of having a choice of pipeline evacuation routes and one of those pipelines is owned in whole or in part by one of more of the JV partners, then it would be an obvious precaution to prevent those conflicted JV partners from voting on the choice of pipeline route. A well constructed JOA would make provision for any such conflict of interest.

However if a partner already has a quantity of oil flowing through one of the competing pipeline candidates from one or more other fields then it may be in its commercial interest to vote for a pipeline route that is not the cheapest one available, even if it does not own a share in that more expensive pipeline. The more costly route may give that particular JV partner the opportunity to achieve economies of scale in its shipping. Or it may provide less transparent optimisation opportunities for that particular JV partner to finesse the number of barrels loaded onto each cargo from each of its fields using that pipeline route. Each of the fields may be derived from different PSCs that are at different stages in their cost recovery cycles or that are subject to different profit share/royalty/tax rates. So which barrels are loaded onto high sales priced cargoes and which barrels go onto cargoes that have achieved a lower price can substantially improve a company's overall economic position in a way that may not be available to its JV partners in any of its fields using that pipeline route.

Does that situation represent a conflict of interest that should prevent that JV partner voting in the choice of pipeline route? That is debatable, but even if it did, it would be very difficult to draft terms in a JOA that would be capable of identifying and addressing the disparate commercial situations and motivations of all the JV partners.

# Value Adjustment Mechanisms

Constructing a whole new pipeline for the sole use of a new oil field can be very costly in terms of both time and money. In some regimes gaining host government consent to do so may be a drawn out process, particularly in countries where a dense population and strong environmental lobby give a wide range of stakeholders a say in the matter. Building a new pipeline is therefore usually only considered when there is a large volume of oil reserves in situ and when the production rate and longevity of production envisaged justifies the cost and effort. In areas where there is no existing pipeline infrastructure this may be the only choice.

Alternatively, when selecting a pipeline out of a range of existing competing options, there are a number of factors that have to be considered. For example:

- Is the pipeline "common carriage" or "contract carriage"?

  - In the case of common carriage the pipeline operator is required, usually by law or by a pipeline regulator, to offer any company wishing to ship oil through the pipeline use of capacity on the same terms as those offered to other pipeline users. If there is a greater volume of oil offered for shipment by pipeline users than the capacity of the pipeline can accommodate then it may be that every pipeline user has to be pro-rationed, i.e. pipeline users are only allowed to ship a proportion of the volumes that they have nominated to deliver;

  - In the case of contract carriage the pipeline operator and the shipper enter into a contract to reserve a proportion of the pipeline capacity for the sole use of a particular field. If the shipper reserves capacity, but is unable to ship because the field production is less than expected, then typically the shipper has to pay for such reserved capacity whether it uses it or not. These are referred to as "send-or-pay" contracts. Contract carriage tariff rates are more likely to vary from shipper to shipper, but these too may be regulated by a pipeline authority;

- What type and rate of tariff applies? These may be based on:

  - The capacity booked, e.g. based on a cubic metres/hour; or,

  - The actual use of the capacity, i.e. throughput based on tonnes or barrels shipped; or,

  - The distance over which the oil is shipped. In long distance pipelines this may be expressed as $/barrel/mile or $/cubic metre/hour/km. In short distance, dense pipeline systems there may be a "postalised" tariff expressed as $ per barrel from point A to point B and there is no commitment that the oil will be shipped by one particular series of pipes rather than another. This is analogous to buying a stamp to get a letter from A to B, regardless of which of the various possible routes to the destination is used.

- Is the oil dispatched in batches or commingled with the oil from other fields that use the same pipeline? In the latter case there may be upper and lower limits on the quality of oil that may be shipped in a particular pipeline to ensure that the value of the whole blended stream is not debased by the acceptance of oil from a particular field that has some adverse quality attribute, such as acid or metals.

Commingled pipelines also typically, but not universally, have a value adjustment mechanism (VAM) that maintains a "value in = value out" principle. In other words when two or more crude oils are commingled for transportation by pipeline the owners want fair value for the blended stream they get back at the end of the pipeline, compared with the single stream they put in at the start of the pipeline. VAM's use comparative crude valuations based on simple products yields multiplied by the respective product prices for the input and output streams[9]. These usually take one of two forms:

- Allocation Mechanisms, which compensate those companies whose oil has been down-graded in the commingling process by giving them more output barrels than input barrels. Companies whose oil has been upgraded by commingling receive fewer output barrels than input barrels.

- Quality Banks, which compensate those companies whose oil has been down-graded in the commingling process by giving them a cash payment. Companies whose oil has been upgraded by commingling are required to make a cash payment.

VAMs are at best a blunt instrument for achieving parity amongst a range of fields all with differing qualities. For example, in valuing oil input and comparing it with the value of oil output, what assumption should be made about the type of refinery that is likely to extract the best value out of a particular type oil? If a simple topping refinery model is used to construct the VAM that does no more than model the separation of crude oil into different broad categories of refined product by simple distillation, then grades of

---

[9] These are called Gross Product Worth calculations and are covered in more detail in Chapter Four and in the companion volume to this book "Trading Refined Products: The Consilience Guide."

crude oil that would yield more value from a cracker or coker or from lubes production will be under-valued by the VAM.

Some quality banks do not even go that far and may compensate shippers only for the differences between the API gravity and sulphur content of crude input compared with the crude output.

Furthermore some quality attributes do not blend in a linear fashion. So introducing a small quantity of a new grade into the commingled blend may bring down the value of the whole blended stream in a way for which a VAM is incapable of compensating.

When a pipeline route is being chosen the VAM may not be negotiable. The pipeline operator will wish to have the same VAM terms with all shippers so that the poor quality fields contributing to the blend are compensating exactly the high quality fields, leaving the pipeline operator neutral. Pipeline operators are unwilling to take on a quality risk that might eat into their tariff income.

## Terminal Logistics

At the end of each pipeline are the terminal storage and collection facilities from which oil is loaded onto oil tankers for sale and delivery ultimately to oil refineries[10]. From the point of view of companies selling oil the more flexible the terminal can be in aggregating daily oil flows into shippable cargo quantities and the wider the range of vessels it can accommodate and the greater the number of jetties it has at which vessels can be loaded, the better the price at which the producers will be able to sell their oil.

But such flexibility costs money in the form of the upfront CAPEX needed to build the terminal. So there has to be a trade-off between what the trading department of a producing company would like to have in an ideal world to maximise the sales price of each cargo and the CAPEX that the asset will bear, or the tariff that will have to be paid to use a third-party pipeline and terminal. That trade-off can only be optimised if those modelling the field economics have access to accurate assessments of what the consequences will be of having less storage capacity or one less jetty etc.

The consequence of 400,000 bbls of storage capacity at a terminal with a pipeline that ships 25,000b/d is obvious - a cargo will have to be shipped out at least every 14-16[11] days to ensure that there is sufficient ullage (i.e. free storage capacity) to avoid shutting in production. What may be less obvious is that crude oil tends to move in the

---

[10] Some crude may be burned directly in boilers or power stations, but the vast majority of oil is sold to refineries for separation, treatment and the upgrading of oil products.

[11] Not all tank stock will be usable so these figures are only rough approximations for illustrative purposes.

international market in tankers that load 600,000 bbls or more. So either the buyer will have to present an uneconomically small tanker to load the cargo or will be required to load a part cargo of 350-400,000 bbls then wait around for up to 8 days to fill up the ship.

The buyer will take this into account when deciding to buy a particular cargo and will deduct its expected "deadfreight" or "demurrage" bills from the price at which it is prepared to buy the oil. Deadfreight is unused tanker carrying capacity that has to be paid for by a vessel charterer whether it used or not. Demurrage is compensation that has to be paid to a vessel owner to compensate for excess "laytime", i.e. time allowed to load oil under the terms of the charterparty – usually 36 hours. Most terminals pay demurrage to vessels that are delayed awaiting cargo, in which case demurrage, in the scenario described above, would be a direct operating cost for the field, rather than being reflected in price under-performance for the field.

Companies building an oil terminal or choosing a pipeline export route can only be encouraged to identify these hidden costs and reflect them fairly in the economic model of the field. Asset team managers may be less than enthusiastic about tracking down such hidden costs, particularly at the development stage of a project. A plea for increased CAPEX or an increased estimate of OPEX may threaten the project's development or may steer the development plan towards an export route that the engineering side of the project does not favour.

By the time the project is on-stream and these costs begin to manifest themselves in reality, the asset team manager may well have moved on to a new project, possibly with a bonus for completing the first project within a strict budget. This leaves the next asset team manager with the job of explaining why the operating costs are so high, or the traders to explain to management, or more likely to the NOC or tax authorities, why the price they achieve when selling the oil is so low.

## FPSOs

In some areas where there is no existing pipeline infrastructure, and where the reserves estimate does not justify constructing a new pipeline, the JV partners in an oil field may opt for a Floating Production and Storage Operation (FPSO). This produces oil direct into either a purpose built tanker or into a converted tanker for stabilisation and storage. Smaller tankers come alongside for ship-to-ship (STS) transfer of oil into vessels that can sail to a wide range of market destinations. It is typically, but not universally, the case that the JV operator of an FPSO will be charged with the responsibility of ensuring that there will be sufficient offtake tankers available that are capable of loading at the field in question. If the FPSO logistics are flexible, the shipping can operate in a similar fashion to that of an onshore terminal. But in hostile waters it is often the case that a particular configuration of tanker is required to perform an STS transfer from an FPSO.

In these circumstances the operator may opt for one of two offtake solutions:

- a long term charter (time charter) of a small number of named vessels with a limited sailing radius that are dedicated to the field; or,

- a contract of affreightment (COA) for a wider range of vessels to perform field-related voyages on request.

The first option is less commonly used today than it was in the 1970s and 1980s because it is costly and it limits the possible market for the crude oil. For example, we will use again for illustrative purposes an FPSO with 400,000 bbls of storage capacity and a production rate of 25,000b/d, but this time we will assume we have one dedicated offtake shuttle tanker with a capacity of, say, 350,000 bbls[12]. Once the shuttle tanker has loaded and sailed it will have up to about 16 days to get to a discharge port unload and sail back to the FPSO in time to load the next cargo. Depending on the location of the FPSO relative to centres of refinery demand that may be a tight operation, where any delays to the vessel by weather or at the discharge port will risk shutting in production. It will almost by definition lock the oil field out of any long-haul arbitrage opportunities. Most of the FPSOs that have opted for dedicated shuttle tanker operations tend to operate with at least two tankers to avoid shut-ins and to widen the market for the oil.

It is more common today for FPSOs to enter into a COA with a large shipping company, such as Teekay or Maersk, who will provide one out of a large pool of vessels suitable for the FPSO loading configuration when needed. This can be more cost effective than a time charter because the JV partners can limit the amount of time that vessels are on charter to the field. If one of the approved vessels is out of operation it is the owner's responsibility to provide a substitute. A COA shares some of the characteristics of having a call option on vessels, but the analogy is not exact because the J.V partners are typically required to pay for a certain minimum number of voyages, whether they use them or not. How high is that minimum number of voyages is subject to negotiation.

---

[12] The J.V. partners may have to pay a higher unit freight cost for such a small tanker: generally the larger the tanker, the lower the daily freight rate. But using a smaller vessel size avoids the deadfreight and demurrage issues referred to above. So a cost-benefit analysis has to be carried out before a chartering decision is taken.

# Lifting Agreements

The lifting agreement (LA) is a key document particularly to the trading and shipping operations staff within an oil producing company. It is the rulebook that allocates individual cargoes of oil to each of the oil companies in the JV partnership and confirms the date and the size of each shipment.

There may be two layers of LAs in operation at a given shared terminal:

- There may be separate LAs signed by the terminal operator and by each field that uses the terminal facilities. Each of these "field LAs" will typically have identical terms, because if there are any conflicts of lifting requirements amongst fields the terminal operator must have a common rulebook for resolving such conflicts.

- Once a cargo of oil is allocated to a particular field by the terminal operator there may be a different LA that is applied by the field operator to allocate the field's cargoes amongst individual JV participants. The field operator determines which of the JV partners is entitled to a cargo that has been allocated to the field by the terminal operator.

In some cases individual companies may sign a terminal LA direct with the terminal operator. This allows a company with interests in multiple fields to a accrue jointly the right to lift cargoes from different oil fields in which it has a partnership interest and that use the same loading terminal.

The timetable for allocating cargoes to companies varies from country to country and from terminal to terminal, but it is usually done on a monthly cycle. Generally, for the lifting month (M), the allocation of cargoes is organised somewhere in the period of 25th M-2 to 10th M-1. So the allocation of cargoes for the lifting month of January will be undertaken from about 25th November to about 10th December, depending on where in the world the field in question is located. Scheduling of cargoes is carried out by the terminal operator based on production forecasts that are made available from the upstream departments of oil producing companies usually in the last 10 days of M-2. So production forecasts for January would be provided to the terminal operator some time between 20th and 30th November. The terminal operator will consider what each field expects to produce in month M and whether that field, or sometimes whether each partner in each field, will have a stock credit or overdraft to take into account in calculating its availability for lifting in cargo-sized quantities in month M. Because oil is produced at a rate of barrels per day (b/d) and is lifted in economic cargoes sizes measured in barrels (bbls) or metric tonnes (mt) fields and companies tend to be over-lifted or under-lifted at any given point in time.

Somewhere before 1st M-1, or around 1st December in the case of the loading month of January, the terminal operator (or field operator), will inform the field operator (or the JV partners) of how much oil is available to that party to lift in month M. The field operator (or each JV partner) informs the terminal operator (or field operator) of its lifting preferences in month M usually around 5th M-1, or 5th December in the case of a January cargo. In some regions this timetable begins much earlier. If there are conflicts with too many companies wanting to lift at a particular time in the month, the operator will attempt to resolve the conflict by discussion. Such conflict of lifting preferences occurs in most months because of the existence of the forward oil price curve introduced in Chapter One. When the market is in contango most producing companies will want to lift their oil later in the more valuable date ranges. In a backwardated market most producing companies will want to lift their oil earlier for the same reason.

If discussion does not succeed in resolving conflicts the LA will determine which field (or company) has priority. This usually involves considering which company has the largest under-lift at the date of lifting, or which company has the greatest time elapsed since it last lifted a cargo, or which company will be least over-lifted after the loading of the cargo in question.

By an iterative process a cargo lifting schedule for month M is produced by the terminal and/or field operator by at least around 10th M-1, or 10th December in the case of the January schedule, in many cases earlier. This is expressed as a series of numbered cargoes arranged into one, two or three – day lifting date ranges with each cargo allocated firmly to an individual company. Once a cargo is allocated to it a company is free to sell that cargo to a third party. Even if production rates do not live up to forecast, the cargo will not be withdrawn from a company, but instead the company will be permitted to over-lift more than expected. That over-lift will be carried forward and worked off by delaying the allocation of further cargoes to the company that is overdrawn in subsequent months (M+1, M+2 etc or February, March etc in the case of January cargoes).

If the cargo lifting schedule for month M (January) is issued on 10th M-1 (10th December), the company that has been allocated the first cargo, say loading in the three-day date range of 1st-3rd M (1st-3rd January), will have only twenty days to organise its sales contract and to find a ship to collect the oil. The company that has been allocated the last cargo, loading in the three-day date range of, say 29th-31st M (29th-31st January), will have about 45 days or more to organise its sale and shipping.

Unless a cargo is sold by about 10 days before the first day of the loading date range, it is considered to be "distressed", particularly if the seller is not an integrated company with its own refinery or storage facility into which unsold oil can be discharged. In these circumstances the seller may have to offer a price discount to potential buyers. Most refineries schedule their next deliveries of crude oil feedstock about 10-20 days in advance, in some regions even earlier, so it may require a price incentive to accept additional cargoes later than this.

## Vessel Nomination and Failure to Lift

Most LAs require producing companies to nominate a vessel (i.e. provide the name of the ship) to perform its lifting, usually not later than 7 days before the first day of the loading date range. If the producing company that is scheduled to lift does not nominate its vessel by then the terminal or field operator, as the case may be, may consider that the company is in a potential "failure to lift" situation and may take steps to lift the oil on the defaulting company's behalf.

Oil is produced into limited storage facilities shared by a number of partners in a field or by a number of fields. Hence if cargoes are not lifted in a timely manner a company that defaults on its lifting obligations runs the risk of shutting in the oil production of its partners and of the other fields that share the same storage facilities. The terminal and/ or field operators will take steps to ensure that does not happen. This may involve selling the cargo on the defaulter's behalf and remitting the proceeds to the defaulting lifter after deducting the operator's costs. The operator may be ill-qualified to make a sale because, as discussed above, the JV operator is unlikely to employ trading personnel, capable of selling a cargo, particularly a distressed cargo, which can be challenging for even the most experienced traders.

Alternatively, the operator may offer the distressed cargo to any other of the JV partners that is capable of lifting it or taking it into its own refinery. The defaulting company may lose the cargo altogether. Or it may only regain its right to the un-lifted oil at the end of the life of the field, which may be several years later.

The costs that can be incurred if a single field, or a number of fields, is, or are, shut-in by a failure to lift can be substantial. These fall on the shoulders of the producer whose ship fails to turn up to lift a cargo. Some may argue that if the producer has sold to a third party and has been let down by his buyer failing to provide a ship, then these costs should be passed on to the third party buyer. However under most industry general terms and conditions of sale the principle of "no consequentials" applies. In other words the costs incurred and profits foregone by producers whose oil has been shut in because one of their number has failed to lift may be considered in law to be too remote to be considered as recoverable damages under a sales contract for one cargo from one producer. This is legal territory and one which should be approached with caution by the layman.

Suffice to say that if oil producing companies are let down by a buyer failing to nominate a lifting vessel, they tend to discount the sales price heavily to find an alternative outlet for the crude oil to ensure that an actual failure to lift does not happen. This is because the cost consequences of shutting in the production of one or more oil fields are likely to be substantial and difficult to measure in advance.

## Physical Sale and Purchase Agreements

The trading of physical oil is typically carried out under term or spot contracts. Term contracts can be for the whole or part of a producing company's forecast or actual production or for the whole or part of a refinery's feedstock requirements; or, it may be for a specified number of cargoes that are purchased by or sold to a single entity for a stipulated period of time. Spot contracts are usually for a single cargo lot, sold usually to a single buyer by a single seller for delivery in a specific date range.

When a buyer and seller agree a deal for the sale and purchase of a cargo, or series of cargoes of crude oil, the contract is typically drafted by the seller. However, if the seller is the smaller of the two companies, drafting rights and obligations can be transferred to the buyer by agreement. The physical oil market is almost by definition an over-the-counter (OTC)[13] one, i.e. a market between a named buyer and a named seller who are entitled to agree any terms to which they both agree and who are each exposed to the other for performance and payment under the contract.

Physical oil contracts typically contained two parts:

- Part One contains the terms specific to the deal in question:

    i.    the counterparties;

    ii.   the grade oil crude oil;

---

[13] This compares with the regulated market conducted on futures exchanges and which will be discussed in detail in Chapter Three.

iii. the type of sale (FOB, CFR, CIF, DAP etc.);

iv. the quantity of the cargo and whether an operational volume tolerance[14] is allowed;

v. the delivery dates;

vi. the delivery location;

vii. the payment terms, particularly whether the deal is done on open credit, or is guaranteed by the parent company, or is covered by an irrevocable documentary letter of credit, or by a stand-by letter of credit, and whether a bank guaranteed letter of indemnity will be produced if the cargo documents are not available when the payment due date arrives; and,

viii. any non-standard terms the parties agree at the time the deal is struck.

- Part Two typically refers to a set of industry standard General Terms and Conditions of sale (GTCs). There are a wide range of GTCs in use in the market each covering broadly similar detailed terms. These usually include:

ix. the precise point at which risk and title passes from the seller to the buyer;

x. shipping requirements and the allocation of responsibility for chartering a vessel, vetting it for its suitability for the loading and/or discharge terminal and seeking its approval by the relevant parties for the particular voyage in question;

xi. sampling and measurement of quality and quantity;

xii. the responsibility for acquiring insurance;

xiii. the particular form that any credit security document might take;

xiv. the applicable law (English? New York? Nigerian? ); and,

xv. a dispute resolution provision.

The GTCs often provide that if a term is not covered in either Part One or Part Two of the contract that "International Chambers of Commerce Rules for the use of domestic

---

[14] This is usually +/- 5% but instances of +/-10% also exist.

and international trade terms" (INCO terms) shall prevail. GTCs should be treated with caution because the standard INCO definitions of types of sales are often silent on issues that are of particular interest to the oil industry, such as whether the delivered quantity is that measured at the load port or the discharge port and where precisely risk and title passes from the seller to the buyer. The GTCs spell out these issues very precisely and sometimes in ways that are counter-intuitive.

For example, in the case of a Free on Board (FOB) sales contract it is unsurprising that risk and title to the oil passes from the seller to the buyer at the load port and that the quantity for which the buyer must pay is the quantity established at the load port, either by the terminal operator or by an independent inspector acting on behalf of both the buyer and the seller. The buyer charters the ship to lift the oil at the loading terminal so it is logical to see that risk and title to the oil, and all that implies for insurance, ocean losses etc., transfers at the loading terminal.

It may well be expected that in the case of a Cost of Insurance and Freight (CIF) sales contract risk and title might pass from the seller to the buyer at the discharge port and that the delivered quantity might be the quantity established at the discharge port. The seller has chartered the vessel and will transfer custody to the buyer at the discharge port. So the transference of risk and title and the measurement and sampling of oil at the discharge port might appear logical. In fact this is not the case. For example consider the "BP Oil International Limited General Terms & Conditions for Sales and Purchases of Crude Oil", a set of GTCs which are widely adopted throughout the industry. Section 8 of these terms makes it clear that the transfer of title and risk and the conduct of measurement and sampling take place *at the loading port* in a CIF contract, except in special circumstances like the delivery of a part cargo. It is only in the case of a Delivered Ex Ship (DES) sale that risk and title and the establishment of quantity and quality occur at the discharge port.

It is unwise to take anything for granted when encountering an unfamiliar set of GTCs in a crude oil sales contract. It is especially unwise to assume that because INCO definitions are applied in one way in, say, a coal or a coffee contract, that the oil industry will apply them in exactly the same way.

There is no substitute for careful checking and confirmation of the meaning of terminology at the time a deal is struck. However as an aid to discussion, Table 2 indicates what is usually meant in the oil industry when a crude oil contract is agreed.

**Table 2 Terminology in Common Usage in the Oil Industry**

(but check the specific contract and GTCs at the time the deal is struck)

| Type of Sale | Free on Board (FOB) | Cost and Freight (CFR or C&F) | Cost Insurance and Freight (CIF) | Delivered Ex Ship (DES)[15] |
|---|---|---|---|---|
| Responsibility for Chartering the Vessel and paying for freight | Buyer | Seller | Seller | Seller |
| Passage of Risk and Title from Seller to Buyer | As the crude oil passes the vessel's permanent hose connection at the loading terminal. | As the crude oil passes the vessel's permanent hose connection at the loading terminal, except in the case of part cargoes where property passes at the discharge port. | As the crude oil passes the vessel's permanent hose connection at the loading terminal, except in the case of part cargoes where property passes at the discharge port. | As the crude oil passes the vessel's permanent hose connection at the discharge port. |
| Measurement and Sampling | At the loading terminal by terminal inspectors or, by agreement, by an independent inspector appointed by both seller and buyer. | At the loading terminal by terminal inspectors or, by agreement, by an independent inspector appointed by both seller and buyer. | At the loading terminal by terminal inspectors or, by agreement, by an independent inspector appointed by both seller and buyer. | At the discharge port by an independent inspector appointed by both seller and buyer. |

---

[15] Replaced in 2010 with DAP, or Delivered at Place

## Table 2 Terminology in Common Usage in the Oil Industry continued

| Type of Sale | Free on Board (FOB) | Cost and Freight (CFR or C&F) | Cost Insurance and Freight (CIF) | Delivered Ex Ship (DES)[15] |
|---|---|---|---|---|
| Vessel Charges | All duties, fees, taxes, quay dues, pilotage, mooring, towage expenses and any other charges due in respect of the vessel at the loading terminal are for the account of the buyer. | All duties, fees, taxes, quay dues, pilotage, mooring, towage expenses and any other charges due in respect of the vessel at the loading terminal are for the account of the seller. All dues and other charges on the seller's vessel at the discharge port, other than those defined by Worldscale[16] as being for the vessel owners' account, shall be borne by the buyer. | All duties, fees, taxes, quay dues, pilotage, mooring, towage expenses and any other charges due in respect of the vessel at the loading terminal are for the account of the seller. All dues and other charges on the seller's vessel at the discharge port, other than those defined by Worldscale as being for the vessel owners' account, shall be borne by the buyer. | All duties, fees, taxes, quay dues, pilotage, mooring, towage expenses and any other charges due in respect of the vessel at the loading terminal are for the account of the seller. All dues and other charges on the seller's vessel at the discharge port shall be borne by the seller. |
| Responsibility for export licence, customs clearance at loading port | Seller | Seller | Seller | Seller |
| Responsibility for obtaining insurance | Buyer | Buyer | Responsibility for obtaining marine insurance for 110% of the cargo value is obtained by the seller. Insurance against other risks such as war, strikes etc. rests with the buyer, but may be transferred to the seller by agreement and payment. | Seller |

---

[16] New Worldwide Tanker Nominal Freight Scale (Worldscale) is an international freight index for tankers, which provides a method of calculation of payment for the transport of oil by ships, for a single or several consecutive voyages. Worldscale is a table giving the amount of dollars per ton for a large number of standard routes.

Table 2 Terminology in Common Usage in the Oil Industry continued

| Type of Sale | Free on Board (FOB) | Cost and Freight (CFR or C&F) | Cost Insurance and Freight (CIF) | Delivered Ex Ship (DES)[15] |
|---|---|---|---|---|
| Responsibility for import licence, customs clearance at discharge port | Buyer | Buyer | Buyer | Buyer, unless the seller is the importer of record. |

# Freight Contracts

Whilst it used to be the case that many large oil companies owned their own ships, this is generally no longer the case. Oil companies now tend to "fix", or charter, tankers from specialist shipping companies. A large percentage of shipping companies are owned by individuals and companies that have no involvement in the oil industry other than owning and offering oil tankers out to charter.

Chartering a ship is typically carried out by contacting a ship broker who will locate the most appropriate vessel for a particular voyage at the most competitive price. Charters can be done on a spot basis for one journey, i.e. voyage charters, or a term basis, i.e. time charters.

In the case of a spot charter, the freight rate is based on moving a specified volume of cargo between two locations or regions at a cost that is typically a function of the size of the cargo. A time charter is a term contract for a named ship that will be made available for an agreed period of time to the charterer to undertake whatever voyages the charterer chooses, within agreed regions. The term can be anything from 3 months to a number of years. The freight charge in a time charter is typically agreed up front as a day rate, and may even be paid up front depending on the relative credit standing on the two counterparties to the deal.

When a suitable ship is identified for a spot or time charter it is usually the owner who makes a firm offer to the charterer. The owner usually leaves that offer on the table for a short period of time, allowing the charterer time to check if the vessel will be accepted by both the cargo loading and discharge terminals. The two parties agree to put the vessel "on subjects", i.e. make a provisional contract, subject to minor details being finally agreed, but still with no firm commitments on either side. The charterer can then check with the likely load port(s) and discharge port(s) whether or not the vessel conforms to the terminal regulations as to length, draft, ballast handling capability etc. Once these checks have been made and satisfactory answers received, the charterparty details can be finalized and the charterparty agreed. The charterparty is the legal contract covering all the terms applicable to the agreement between ship owner and charterer.

An alternative type of ship charterparty is the Contract of Affreightment (COA). A COA is a contract not for a specific ship, but for the owner to provide a suitable ship out of a pool of qualifying vessels at the required load port for delivery to the destination at a series of times nominated regularly by the charterer.

## Freight Costs

In the case of a spot charter, the freight rate agreed may be a price per unit of cargo, for example $X or $Y per tonne of oil, or a lump sum of $Z for voyage from Port A to Port B. The size of the cargo will be agreed in advance, usually with a tolerance of between 5 and 10%. This volume tolerance is typically described as "more or less at owner's option" (MOLOO).

For time charters the ship-owner and the charterer normally agree a rate per day or perhaps per dead weight tonne (DWT)[17] per day.

For single voyage charters freight is normally calculated by means of a system known as Worldscale. The "New Worldwide Tanker Nominal Freight Scale," or Worldscale, was devised during World War II to compensate ship owners for the use of their ships by the navy. It is an international freight index for tankers, which provides a method of calculating the cost of transporting oil by a reference ship at a reference speed under reference sea conditions for a single voyage between two ports. This is updated annually.

The Worldscale on-line book provides a table giving an amount in dollars per tonne for every likely combination of origin and destination ports for oil. This amount is based on a standard size ship of 75,000 tonnes dead weight, for a round trip voyage via the shortest route travelling at 14.5 knots and using 55 tonnes of bunker fuel per day during the voyage (a different amount applies whilst at the port). It assumes average weather conditions, taxes, and a standard return on capital. These rates are known as Worldscale flat rates or Worldscale 100 or WS100, and are revised annually in the light of changing conditions, particularly varying with the cost of bunker fuel.

---

[17] DWT is a measurement of the weight that the ship can transport which includes cargo, fuel, supplies, and crew.

Owners and charterers, possibly via ship brokers, negotiate a freight rate, which is a proportion of the Worldscale flat rate for a specific voyage. This price diverges from Worldscale 100 depending on how much the specific charter diverges from the reference conditions under which Worldscale 100 is calculated, the vessel size and type, the rising and falling costs of fuel and the supply and demand for this type of vessel at any given time. Generally the smaller the ship the higher the Worldscale rate because of economies of scale. The proportion of Worldscale 100 that is agreed on any given day is tracked by brokers and published in various sources such as Argus, Platts and The Baltic Exchange etc.

The Baltic Exchange in London was founded in 1744 and originally matched ships with cargoes of tallow from the Baltic region, hence the derivation of its name. It provides data on shipping and freight rates and reports freight transactions. The Baltic international tanker routes (BITR) reports on a number of the most popular international routes each day at a given time, and produces a Baltic Clean Tanker Index (BCTI) and the Baltic Dirty Tanker Index (BDTI). These routes and the type of tankers are very specific.

So, for example, "BDTI 1" reports the market rate for a 280,000 mt vessel on a voyage from the Middle East Gulf to the US Gulf, specifically "Ras Tanura to the Louisiana Offshore Oil Port (LOOP), with laydays (loading date range)/cancelling twenty/thirty days" in advance of the date of fixing the tanker for a vessel with a maximum age of twenty years. "Laydays/Cancelling twenty/thirty" means that the vessel should be available to load within 20 days and if it is not available by 30 days from the chartering date the charterer is free to fix an alternative vessel and commence litigation against the defaulting owner.

These quoted rates are now used as references by charterers, shipping companies and trading companies as a hedging and speculative tool in freight forward agreements (FFAs). In this way, for example, an oil company can hedge a major element of the cost of moving oil internationally and lock in a margin for a locational arbitrage.

Once a charterer and owner have agreed the Worldscale rate, say WS80, it can look up the flat rate provided by Worldscale for the actual loading and discharge ports, say $1.20/mt, and multiply this by the market Worldscale rate expressed as a percentage. So in this case WS80 means 80% of $1.20/mt = $0.96/mt and that is the cost of the voyage (Note sometimes there are additional costs that must be added for certain routes, such as canal fees, or for certain ports charges and taxes).

The contract specifies the method by which the freight payment is calculated. This will include agreeing how the weight of cargo used in the freight payment calculation relates to the actual cargo size. For example, if the ship can take 75,000 mt but the cargo size is only 65,000 mt there is potentially 10,000 mt of spare capacity or "dead freight" on the vessel. Who pays for the deadfreight is spelled out in the charterparty.

## Laytime and Demurrage

The time allowed in the charterparty for the cargo to load and discharge is referred to as "laytime" and may be, for example, 36 running hours. The length of the **laytime** is dependent on the size and pumping rate of the ship and the size of cargo. If the cargo takes longer than the laytime allowed to load and/or discharge then additional charges in the form of "demurrage" may have to be paid by the charterer to the owner. Depending on the circumstances under which the ship has taken more time than the laytime allowed to load and/or discharge, then the charterer may be able to pass on this demurrage claim from the owner to the loading or discharge terminal. Alternatively if the reason the loading took longer than the laytime was for the ship's purposes, then the charterer may be able to rebuff the demurrage claim from the owner.

The demurrage clause in the charterparty spells out when laytime actually starts, i.e. when the "time allowed" clock starts ticking. Often this is at commencement of loading, or, regardless of whether loading has commenced, 6 hours after the ship's captain tenders Notice of Readiness (NOR) to the port. This is a formal notification that the vessel is ready to come onto the jetty to load or discharge.

The NOR is required to be given by the vessel to the terminal during a specified date range referred to in the charterparty as the **laydays.** The charterer will agree laydays that are back-to-back with the delivery date range it has agreed under its crude oil sale and purchase contract.

The charterer is not obliged to start loading before the laydays agreed with the terminal commence. If NOR is tendered before the laydays and the terminal does not agree that the vessel can come in early, then NOR is deemed to have been given at 00.01 hours on the first day of the agreed laydays or on commencement of loading (or discharge), whichever is the earlier. Laytime, usually 36 hours as mentioned above, commences 6 hours later, allowing time for inward passage of the ship.

If the NOR is tendered after the laydays agreed with the terminal have passed, then the terminal can delay accepting the ship until it has accommodated those vessels that have arrived within their proper laydays, even if those laydays are later. In these circumstances laytime starts to run on the actual commencement of loading (or discharge).

Laytime usually finishes when the loading (or discharging) arms are disconnected from the ship. However, time identified in the terminal time sheet as having been used for particular purposes may be excluded from the laytime allowed. For example delays caused by bad weather or for waiting for a pilot or tugs may be excluded from the laytime calculation. If the port or terminal delay a vessel for longer than the agreed laytime under the physical oil sale or purchase agreement, the terminal will be liable to pay demurrage to the charterer of the vessel. Charterers seek to ensure that the demurrage terms under their physical oil contracts are back-to-back with the terms under which the charterer has to pay demurrage to the vessel owner.

Demurrage claims against a loading or discharge terminal are usually subject to a "time bar", i.e. a claim for excess time must be submitted within, say, 30 days or 90 days, or the claim is null and void. So the cost conscious trading company deals with demurrage claims promptly to ensure that they have passed them on to the terminal from the owner before the time bar comes into force. The party holding the demurrage claim "parcel" when the music stops, i.e. when the time bar clicks in, ends up bearing the cost.

Usually when a trading company charters a ship, whilst it is likely that it knows where the cargo will load, it may not yet have made a corresponding sale. Even if it has, the company to which it has sold may not have made a sale itself so at the time of the tanker charter the discharge port may be unknown. Therefore a charterparty may often not give the final discharge port, but will include different rates for different possible regions of discharge. Traders seek to obtain the maximum flexibility they can in the load and discharge ports that they are entitled to visit. This allows them to find the best priced sales outlet without being hindered by the fact that the ship can only discharge in a limited region or at a limited number of ports. After the vessel is chartered, but before it loads, the trader may wish to deploy it on an entirely different voyage from a different load port where a new profit opportunity has arisen. It is not unusual for a cargo to load and sail without a final destination being known.

This Chapter has considered a cascade of agreements whereby crude oil is discovered, developed, produced, transported, stored, accumulated into cargo lots and allocated to individual companies for onward sale and shipment to traders and ultimately to refineries. There are several "trading" messages to take away from this Chapter:

1. The first oil price risk that an oil company faces is that the price at which it is taxed, recovers costs and distributes profit share upstream may vary substantially from the

price at which it sells its oil in the market, regardless of the fact that the PSC may appear to say differently;

2. CAPEX savings in the field design will have consequences for the trading flexibility, operating costs and the price of oil in the market. Producers would be well-advised to consider and evaluate appropriately these factors in the economic model of the field;

3. The choice of transportation route for exporting the oil requires careful evaluation not only of capital and tariff costs but of the value in/value out adjustment mechanism for compensating producers contributing oil of varying quality to a commingled blend. There may be hidden conflicts of interest amongst JV partners in the choice of export route;

4. The price of a cargo of oil varies quite substantially with its delivery date. The JOA ought to protect partners against the cherry-picking of choice dates by the operator of the lifting agreement. However if minority partners sell their production to the field operating partner the actual allocation and sale of cargoes may be conducted on the other side of the operator's internal Chinese Wall;

5. The sales contract often does not reflect a fair share of the additional value that the operating company can extract from organising cargo liftings in a tax-efficient manner and to optimise its own operational considerations. In other cases the traders of the operating company may agree to pay a premium over the ostensible market value of the oil, particularly if there is competition for cargoes from third party buyers. Small producers often do not fully appreciate why a premium is paid at all and therefore are not best-placed to assess if the level of the premium is appropriate to the circumstances of the field; and,

6. The avoidance of any "failure to lift" situation is of over-riding importance to an upstream producer. The consequences of shutting in the production of JV partners or partners in other commingled fields are large, difficult to quantify and may not be capable of being passed on to a defaulting third party.

The next Chapter will begin the process of unbundling the oil price into its component parts and will examine the absolute price, A, in detail.

# CHAPTER 3

❝ Always keep tight hold of nurse,
For fear of finding something worse ❞

*Hilaire Belloc*

## Introduction

The previous Chapter of this book described the cascade of upstream agreements whereby oil is produced, transported and allocated to oil producing companies for sale to the market under "physical" crude oil contracts at formula prices. This Chapter will start to look at the detailed anatomy of oil prices beginning with the absolute oil price component, A.

"A" represents the general level of oil prices and it is determined over time by macroeconomic factors such as international economic performance, oil production levels, geo-political factors, environmental regulation etc.

In Chapter One under the heading "The Three Components of the Oil Price Formula" we introduced the concept of the forward oil price curve and said that the price of any given cargo of crude oil is typically made up of three different components. These are:

• 　　　the Absolute Price (A);

• 　　　a Time Differential (T); and,

• 　　　a Grade Differential (G).

We said that the forward oil price curve is not a forecast of what prices will be at some future date. Instead it is a snapshot taken at a particular instant in time of the prices at which buyers and sellers are actually prepared to deal at that moment in oil for delivery at different dates in the future. If a buyer or a seller has a financial imperative that requires it to lock in the future sale or purchase price of oil in, say, 6 months, a year or 5 years forward, it can do so using the range of physical and financial instruments described in this book.

A company may wish to enter into a trade up to 5 or more years forward for the purposes of under-writing a base case price assumption in its 5-year scenario plan. Or it may wish to do so to guarantee a loan taken out to finance the development of a particular asset. Or it may wish to simply back its own oil price view on a purely speculative basis. There is no universal right or wrong reason for entering into trades based on the forward oil curve, so long as those trades are appropriate to the memorandum and articles of association of the company concerned and are properly authorized and suitably financed by a board delegated mandate.

If the question were to be asked "what is the price of oil today?" the answer would be slightly different depending on who was asking and where they were physically located. If the question were to be asked in the US, the answer would most likely be couched in terms of "WTI". If the question was asked in Europe, Africa or the Mediterranean the answer might be couched in terms of Brent. In the Middle East or Japan the answer might be couched in terms of Dubai. All three answers would be right, but would be very different because each would be referring to a different benchmark grade.

If the same question were to be asked of a trader – "what is the price of oil today?" – in order to answer the question the trader would require considerably more detail. The trader would ask "Which benchmark (A), for delivery in which time period (T) and for determining the price of which particular grade of oil (G)?" It is the task of this Chapter to address the value of A. The value and characteristics of T and G will be explored in detail in subsequent Chapters.

The absolute level of price, A, is the most volatile of the three components of the oil price and this component is routinely hedged by oil producing and refining companies. It is the price of deals done in the futures or forwards markets, expressed in terms of one of the benchmark grades.

## What is a Benchmark?

In Chapter One we explained that the majority of physical oil cargoes are sold and priced by reference to a formula, rather than expressed as a fixed and flat price of $X/bbl. We also explained that this formula often refers to the average price of a benchmark grade,

A, as published by a price assessment or reporting agency either during the month of delivery in the case of term contracts, or on the 3-5 days around or just after the anticipated bill of lading (B/L) date in the case of spot contracts. The formula adjusts the average published value of A to reflect the time differential, T, and the grade differential, G, relevant to the specific cargo in question. Differentials are by definition an adjustment to an absolute amount so in order for a formula to be resolved the value of the absolute amount has ultimately to be determined in terms of dollars and cents per barrel.

One of the key prerequisites of a successful benchmark grade of oil is the existence of a market in that grade expressed in $/bbl, not by reference to a formula. Hence the most active crude oil benchmarks are those where there is liquid[18] trading in a "flat-priced", i.e. $/bbl, forward or futures contract. Other key characteristics of the ideal benchmark grade would be:

- A large volume of production, such that it is difficult for any party to "corner the market";

- A large number of producers to prevent one company, whether a NOC, an oil major or a large independent, controlling supply;

- Stable quality that does not have any particularly difficult physical attributes, so that the grade can be bought by a large number of refiners;

- Good loading terminal logistics with enough storage to accommodate a number of days of production with sufficient flexibility to handle operational changes and shipping delays;

- Sufficient jetties with capacity to load a range of tankers to optimise freight and promote inter-regional arbitrage;

- A transparent lifting schedule so that all buyers and sellers can assess the changing availability of cargoes on an equal footing;

- Standardised, transparent general terms and conditions of trade (GTCs), so that companies can buy and sell repeatedly on back-to-back terms; and,

- A benign host government that does not intervene in either price or supply.

There are several grades around the world that are used as "markers" or benchmarks for their own area. None of these benchmarks tick all the boxes needed in order to be considered an ideal benchmark. There are no new ones on the horizon with better

---

[18] Liquidity refers to the turnover in the market. The more liquid a contract the more actively it is traded.

prospects of being the universal benchmark of choice by the industry. Arguably, the existing benchmarks are:

- Dubai

- ESPO

- WTI

- ASCI

- Tapis

- Brent

## Dubai

Dubai is a medium light (~30.5°API) and high sulphur crude oil (~2.0 %), which is typical of the region. In the past Dubai was the sole Middle East benchmark grade, because it was one of the few where the government did not frown on spot market trading. However its production has declined to less than 50,000b/d calling into question its long-term prospects as a benchmark grade. To bolster the volume the alternative grades, Oman and Lower Zakum, were introduced for delivery into the Dubai contract and other substitute grades may be included in future.

It is has traded both as a physical contract and in a forward market, although the forward market is now so thin as to be almost non-existent. It now trades as a very liquid swap contract[19]. The swap market trades in small partial cargoes that can be built up into a deliverable quantity or settled in cash by entering into an equal and opposite position with the same counterparty with whom a position has been opened.

Perhaps the most significant threat to Dubai's status as an oil price marker was the Dubai government's decision to reclaim direct control of its petroleum assets in 2007, on the expiry of an offshore oil concession that had been held by a ConocoPhillips operated consortium for 40 years.

## ESPO

The comparatively new East Siberian Pacific Ocean (ESPO) pipeline transports a medium gravity (34.5-35° API), medium to low sulphur (0.5%) grade of crude oil from the East Siberian producing area around Taishet to the Port of Kozmino on the Pacific Coast. The first delivery was in December 2009. The pipeline was completed in December 2012,

---

[19] Swap contracts will be discussed in detail in Chapter Six.

but before then crude oil was transported by rail car on the final leg of its journey from Skovorodino to Kozmino. Total pipeline capacity is forecast to rise to 1.6 millionb/d at a later date.

A spur transports 300,000b/d straight to Daqing in China, partly to feed the newly expanded refinery at that location. The Chinese government is seeking to expand this capacity and supply contract, but is meeting resistance from Russia.

Various explanations have been given for this reluctance: the commencement of Urals exports through the new Gulf of Finland port of Ust Luga and increased deliveries through Primorsk on the Baltic, at the expense of Black Sea deliveries, taken together with increased deliveries to the East has given Russia a surplus export capability.

The prospect of ESPO evolving into a new benchmark grade of crude oil has been mooted and certainly it ticks some of the boxes necessary to take on that role over time. But in its early life ESPO is not trading in fixed and flat price contracts, but is itself priced by reference the Dubai benchmark and occasionally to the Brent benchmark. That could change very quickly now that the pipeline is complete and more traders come into the market. It could be that we will see a forward contract in ESPO developing. However the unpredictable role of the state in export and pricing policy may well rule out the use of ESPO as a benchmark.

**Figure 14 The East Siberian Pacific Ocean Pipeline**

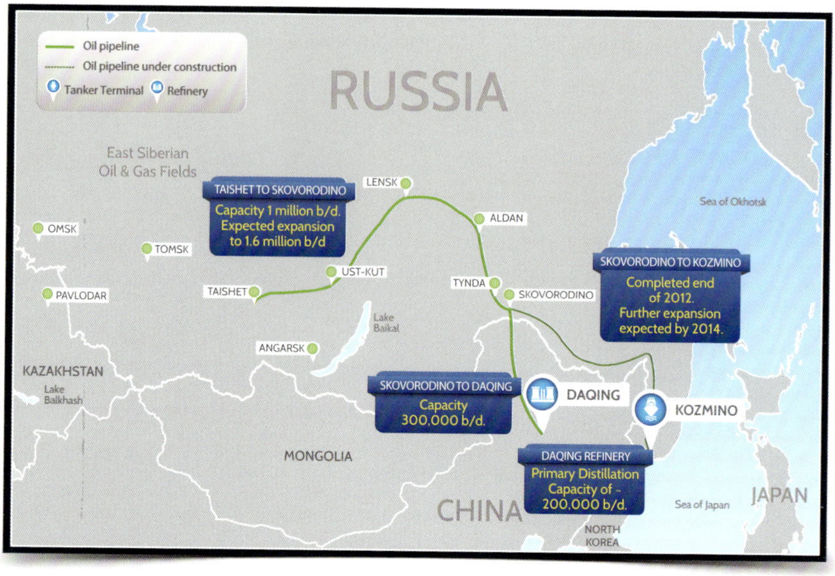

# "WTI"

Initially in the early 1980s after the decontrol of oil prices by the US government, West Texas Intermediate (WTI) crude began to be used as the reference point, or benchmark, for establishing the absolute price of oil, A, internationally. WTI is a US domestic blended grade of light sweet crude oils that is sold by independent producers to buyers, or aggregators, at the wellhead at posted prices. As the name suggests the oil was originally produced in West Texas. It has a typical quality of 38-40 °API and an approximate sulphur content of 0.3%.

The New York Mercantile Exchange's (NYMEX)[20] "WTI" futures contract pre-dated its International Petroleum Exchange (IPE)[21] equivalent contract, based on Brent, by 5 years, commencing trading in March 1983. Although it is commonly referred to as the WTI contract, the contract is more properly called "Light Sweet Crude Oil" and it represents crude delivered at Cushing, Oklahoma. The actual oil traded or delivered can be WTI or five other US domestic grades or UK Brent or Forties or Norwegian Oseberg or Nigerian Bonny Light or Qua Iboe or Colombian Cusiana.

Because WTI used to be the most actively traded grade of oil in the world and its price reflected the supply and demand balance in the world's largest consuming country, its price has been influential as a price reference point for other international grades of crude oil for many years. More recently, in the last 5 years, WTI has suffered from logistical bottlenecks in the US system, particularly at Cushing, the delivery point for the WTI contract.

Oil gathered from producers passes through Midland, Texas from where it can either go south to the Gulf Coast refineries or north to Cushing. Once it gets to Cushing it could only go north through the Spearhead pipeline to refineries in the Chicago area or through the Ozark pipeline to Wood River and it cannot be diverted to the Gulf Coast refineries. As oil production has built up in North Dakota from the prolific Bakken Shale and as imports from Canada down the Keystone pipeline to Cushing have come on-stream, storage capacity and accessible refining demand at Cushing has proved inadequate to mop up supply. Consequently the price of WTI has diverged from the rest of the US market and from that of international sea-borne oil. After trading at a $1-2/bbl premium for many years the price of WTI fell to almost $27/bbl under Brent before recovering to a discount in the high teens of dollars.

---

[20] Taken over by the Chicago Mercantile Exchange (CME) in 2008
[21] Taken over by the Intercontinental Exchange (ICE) in 2001

Unsurprisingly this has provided an economic incentive to truck oil out of Cushing at an estimated $10-12/bbl cost, but has also unleashed plans to improve pipeline infrastructure. For example in May 2012 the reversal of the 150,000b/d Seaway pipeline previously from Freeport, Texas to Cushing, Oklahoma was completed. A further increase in capacity on this line to 400,000b/d is anticipated in early 2013 and there are plans to reach 850,000b/d in 2014. But as domestic production of shale oil continues to grow rapidly the need for new infrastructure is likely to be playing catch up for the next few years.

However major plans to extend the Keystone pipeline to bring Canadian crude all the way through the US beyond Cushing to the Gulf Coast where it could be exported into the international market were vetoed by Barak Obama, a decision that is being re-visited following the November 2012 election. Canadian complaints at the comparatively low price it is receiving for its oil in the US domestic market are becoming quite strident.

An approximate $20/bbl disconnect between the inland US domestic market and the international market is unsustainable and is sufficiently large to incentivise corrective investment. If the US government were to continue to block pipeline plans to correct the disparity it would be predictable to see Canadian oil supplies to America decline. The Enbridge Northern Gateway pipeline project to export Alberta tar sand oil across British Colombia to the Pacific is one likely investment, not under US control, that would make this happen.

From a trading perspective the relationship between WTI and the rest of the market in the future is extremely difficult to predict. Physical cargo traders and hedgers outside the US would be ill-advised to use WTI as a benchmark reference point in crude oil contracts until this situation clarifies, which may take several years. This is particularly the case for long term strategic hedges and oil project financing deals of more than a year's duration. Watch this space for the emergence of a new US Gulf Coast fixed and flat benchmark in the coming years.

**Figure 15 Major US Canadian Crude Pipelines (approximate schematic)**

## ASCI

While WTI has been dancing to its own tune out of step with the rest of the market, a new quasi benchmark has emerged- the Argus Sour Crude Index (ASCI). This has been published by Argus Media since May 2009. It represents a volume-weighted average of deals in the US Gulf coast medium sour crudes Mars, Poseidon and Southern Green Canyon (SGC). The characteristics of these grades are as follows:

• Mars: 28.9°API and 1.93% Sulphur, sold FOB Clovelly, Louisiana

• Poseidon: 30.9°API, 1.72% Sulphur, sold FOB Houma, Louisiana

• Southern Green Canyon: 28.7°API, 2.36 % Sulphur, sold FOB Nederland or Texas City, Texas.

The ASCI price is expressed as a differential to WTI, causing some to question whether or not it is a true benchmark. The ASCI/WTI differential adjusts regularly as the price of WTI oscillates, leaving ASCI with a direct price connection to the international market. This distinguishes the ASCI/WTI differential from what we will discuss in Chapter Five, a grade differential, G. In the case of a typical grade differential the relationship between the grade in question and its benchmark is comparatively stable and it usually only varies if there is a quality or logistical event effecting the non-benchmark grade. The ASCI/WTI differential usually varies to compensate for changes in the benchmark WTI logistics.

Regardless of what it is called, benchmark or grade differential, the ASCI price quotation has attracted some powerful supporters. On 28th October 2009 Saudi Aramco announced that it would begin using ASCI as the benchmark price for all grades of crude oil sold to US customers from January 2010. In December 2009 the state-run Kuwait Petroleum Corporation announced that it would also begin to price its US bound oil by reference to ASCI from January 2010. In February 2010 Iraq's State Oil Marketing Organization (SOMO) followed suit and announced that from April 2010 it too would price the oil it sent to the US by reference to ASCI.

## TAPIS

Despite its status as an Asian benchmark for more than 20 years, Tapis is of little more than historic interest now. Produced in Malaysia, Tapis is a very light (46°API) and very low sulphur crude oil (<0.03%), typical of the region. In its heyday it traded both as a physical grade and a forward market, but the forward market has now died almost completely to be replaced by a partial swaps[22] market. Its production level is now less than 200,000b/d, although this should rise substantially by 2013 after major investment by Exxon Mobil on enhanced oil recovery.

The final nail in Tapis' coffin as a benchmark came in June 2011 when Malaysian state oil company, Petronas, abandoned the use of a Tapis fixed and flat price, sourced from the mechanistic publication, Asian Petroleum Price Index (APPI), in favour of a Brent benchmark in setting its Official Selling Price (OSP). Which brings us neatly to the subject of Brent.

## Brent

The grade of oil commonly described as "Brent" is currently the most actively traded international price benchmark. Despite the fact that Brent is a flawed marker, as discussed below, it continues to be the most influential benchmark and one which continues to be adopted by an increasing number of oil companies, NOCs and fiscal authorities worldwide.

Claims have been made by interested parties that Brent is used to set the value of the absolute price, A, in about two thirds of the world's crude oil contracts. The exact number is difficult to confirm, but Brent is certainly the most pervasive price reference point used by the physical oil market. When the futures, swaps and options markets are taken into account the volume of oil priced by reference to Brent every day is undoubtedly many multiples of daily crude oil production worldwide.

---

[22] Swap contracts will be discussed in detail in Chapter Six.

We said in Figure 8 that daily worldwide crude oil production was, according to BP, about 83.6 millionb/d in 2011. Let's say that estimates are right and that two thirds of this physical volume, about 55 millionb/d or 20 billion bbls per year, of physical oil cargoes is priced by reference to Brent. The ICE futures contract for Brent traded >147 billion bbls in the year of 2012 (see Figure 16) and we can only guess at the size of the OTC forward and derivatives markets. But it is probably being unduly conservative to estimate that Brent is a reference point for at least 200 billion bbls per year in all types of contracts- physical, regulated futures and OTC derivatives. The number is very probably considerably higher. This is before we consider the number of European gas contracts that are priced by reference to oil, in some cases Brent crude oil.

**Figure 16 ICE Brent Futures Volume**

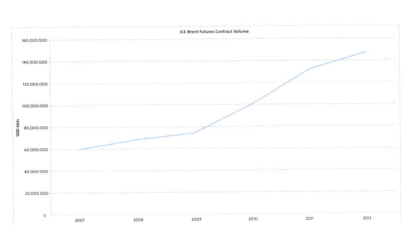

*Source: ICE*

With the pricing of probably about 200+ billion bbls per year based on Brent, it is surprising to see that the "Brent market" is built on a fast reducing physical production foundation. Furthermore only a few of the actual transactions in cargoes of Brent are considered in the database used to establish the reported price of Brent each day, as discussed below.

**Figure 17 The Physical Base of Brent**

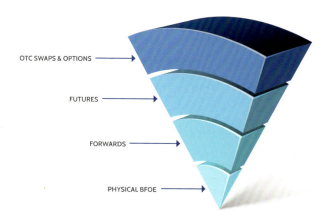

What is known as **Brent Blend** is a stream of oil commingled from more than twenty crude oil and condensate fields exported from the Sullom Voe oil terminal in the Shetland Isles, off the north coast of Scotland. The quality of the blend is about 37-38 °API, with a sulphur content of 0.4%. Average production in 2011 was 146,000b/d, equivalent to about 7 full cargoes, each of 600,000 bbls, per month. This is the physical market in "wet" or "Dated" cargoes that are actually loaded into tankers.

The timetable for establishing which oil producer is entitled to lift each of these about 7 physical cargoes each month was discussed in Chapter Two. To recap, the allocation of available cargo parcels to producing companies for any Month (M) is established in the lifting programme that is issued around 5-10th M-1. In the case of Brent the terminal lifting operator attempts to issue the programme by 5th M-1 for reasons that will become clear when we discuss the 25-Day BFOE market below.

**Figure 18 Brent, Forties, Oseberg and Ekofisk Pipelines**

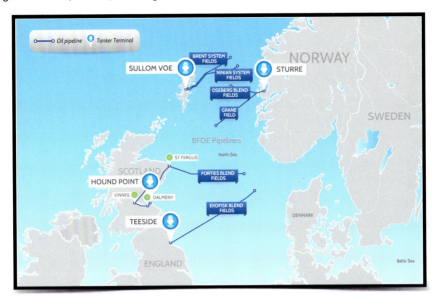

In fact what is called "Dated Brent", or North Sea Dated, by the price reporting agencies (PRAs) is actually a basket of crudes including Brent Blend, Forties Blend, Oseberg Blend and Ekofisk Blend. The grade that determines the actual value of "Dated Brent" as published by the PRAs at any point in time is the grade that is the *least* valuable of the four at that same point in time. This is because the limited volume of Brent means that there are insufficient transactions to produce a reliable assessment of its price. So transactions in these other three grades are also considered to bulk up the database.

**Forties Blend** is the largest North Sea export blend. It is loaded from Hound Point on the east coast of Scotland. Average production in 2011 was about 433,000 b/d amounting to about 22 full cargoes, each of 600,000 bbls, per month. Forties Blend comprises production from over 55 individual fields and is delivered to shore by pipeline. The quality was historically 44.6° API, 0.19% sulphur, but this changed markedly with introduction of the high sulphur Buzzard oil field into Forties Blend in early 2007. Buzzard is a medium gravity, sour crude with an API of 32.6° and a whole crude sulphur level of 1.44%. Consequently the API gravity of Forties Blend is frequently below 40° and the sulphur can be >0.6% because it varies depending on how much Buzzard is being produced relative to production from the other Forties Blend fields at any given time.

Table 3 shows an estimate of the variation in Forties Blend quality when it contains different proportions of Buzzard.

Table 3 The Impact of Buzzard on Forties Blend

| Buzzard % vol | Forties Blend API | Forties Blend Sulphur (%wt) |
|---|---|---|
| 0% | 44.1 | 0.20% |
| 5% | 43.6 | 0.27% |
| 10% | 43 | 0.33% |
| 15% | 42.4 | 0.40% |
| 20% | 41.8 | 0.46% |
| 25% | 41.2 | 0.53% |
| 30% | 40.6 | 0.59% |
| 35% | 40 | 0.65% |
| 40% | 39.4 | 0.72% |
| 45% | 38.8 | 0.78% |
| 50% | 38.3 | 0.84% |
| 55% | 37.7 | 0.90% |
| 60% | 37.1 | 0.96% |

*Source: BP*

Consequently a quality price de-escalator had to be introduced to cope with the higher and variable sulphur content of Forties Blend. This makes Forties Blend less like the Brent Blend grade it is used to supplement in the Dated Brent assessment process. For every 0.10% of sulphur content by which an individual cargo exceeds a limit of 0.6% sulphur, the price of the cargo is de-escalated by a number of cents/bbl that is revised by the price reporting agency (PRA), Platts, periodically to reflect the differential between high and low sulphur crude and products in the market. The size of this de-escalator has varied between $0.20/bbl and $0.60/bbl[23].

**Oseberg Blend** is the main FOB export blend from the Norwegian Sector. It includes production from 11 fields closely located around the Oseberg Field and also limited quantities of the high acid Grane field. Average production in 2011 was 150,000b/d, which amounts to about 8 full cargoes, each of 600,000 bbls, per month. The quality is about 38-38.5°API and the sulphur content about 0.20-0.25 %. Oil is delivered by pipeline to the terminal at Sture in western Norway.

---

[23] At the time of writing this is $0.35/bbl.

**Ekofisk Blend** is second largest FOB export blend from the UK/Norwegian Sector. It comprises production from many fields located in both the UK and Norwegian sectors. Average production in 2011 was 311,000b/d, which amounts to about 16 full cargoes, each of 600,000 bbls, per month. The quality is about 37.5-37.8°API, and the sulphur content about 0.2-0.3 %. Oil is delivered by pipeline to the terminal at Teesside in UK.

The precise quality of any individual cargo of each of the blends varies depending on the production levels of all of the fields feeding into the blend at any given point in time. This can produce unexpected changes when particular fields shut down with operational problems. In summer when large scale field maintenance shut-downs are underway it is necessary for buyers to track potential quality variations closely.

Larger cargoes than 600,000 bbls can be loaded at the four terminals, but the lifting programme tends to be organised in 600,000 bbl parcels. This is because 600,000 bbls is what the oil producers request when nominating cargoes to the terminal operators. Cargoes of 600,000 bbls give producers the flexibility to sell the oil in the Dated physical market or to sell it into the 25-Day BFOE forward market, which trades in 600,000 bbl cargo lots. We will discuss the forward market later in this Chapter.

From the moment the schedule of liftings for month, M, is issued by the terminal operators on or before 5th-10th M-1, producers can sell cargoes into the "Dated" market. So on 10th M-1 a producer could sell a cargo loading in the three day date range of, say, 29-31st M. However in the case of Brent, Forties, Oseberg and Ekofisk, producers that are not locked into term contracts tend only to sell their cargoes in spot contracts that identify the date range of the cargo being delivered if that delivery date range is less than 25 days forward. This is because of the alternative sales option that the 25-Day BFOE market affords.

## Changes Needed to the Brent Benchmark

During 2012 the physical production underlying the Brent benchmark has declined further and work is already underway to decide how best to shore it up.

On 29th June 2007, Energy Compass published an article entitled "Rethinking the North Sea Trade" that reported Consilience's [CEAG's] suggested solution at that time. It said "… CEAG instead proposed that the industry agree a standard Brent reference quality with a starting point for further discussion, somewhere around 38°API, 0.4% sulphur, plus a simple, transparent escalation/de-escalation formula based on API and sulphur that applies to all grades delivered into the 21-Day contract [now 25-Day contract] – not just Forties…...the contractual price for a 21-Day BFOE [now 25-Day BFOE] could then be agreed at the standard Brent reference quality, with the final invoice adjusted for the bill of lading (B/L) quality of each cargo."

At the time this suggestion was rejected because it is very difficult to establish an escalator/de-escalator that will suit everyone: each refiner that processes a grade of oil will see the value of its quality attributes very differently depending on the capability of its refinery equipment. Even the level of the simple Forties de-escalator that was introduced in 2007, discussed above, is a frequent source of uncertainty and dispute in the oil trading community.

Nevertheless a very similar idea was revived by Platts at an industry forum in late 2012 as a possible way of handling the introduction of more international and increasingly disparate qualities of oil into the BFOE basket to boost the underlying volume of the benchmark. The issue is one that will require careful consideration by the industry in the coming months and years. In Consilience's opinion escalators/de-escalators are still not a great idea because of the same problem referred to above- each refinery has a different view of the impact of quality on the value of a grade of crude oil.

On 8th February 2013 Shell, rapidly followed by BP, announced that from May 2013 it would introduce a range of escalators to apply when grades other than Forties are delivered into the 25-Day BFOE forward contract. This is discussed in Appendix 1 to this book. The setting of escalators and de-escalators might best be agreed and coordinated by a representative group of all those stakeholders on whom it will have an impact. This might include the taxation authorities and PSC interests that have an exposure to those grades of oil included in any new basket.

In Consilience's opinion an expert panel, subject to regulatory oversight, might be convened to deal with contractual trading housekeeping issues of this type in the oil market. This panel might include representatives of major and independent producing/ refining/ trading companies/regulated exchanges/ NOCs/PRAs and fiscal authorities. The terms of reference of such a panel might be to host regular industry meetings at which issues of relevance to oil trading contracts might be discussed. These issues might include declining volumes of key benchmarks, quality issues, logistical problems etc. that might require a concerted contractual amendment by stakeholders or a change in reporting methodologies of the PRAs. Regulatory over-sight would probably be necessary so that companies can talk freely without fear of anti-trust allegations. The introduction of a Brent reference quality with escalators and de-escalators applied to the price when grades of different qualities are delivered would be a classic example of an issue that might be considered by any such panel.

# The Regulation of Benchmarks

Given the influence of the Brent price benchmark, the price setting and reporting methodology has to be a robust and a reliable reflection of true market price levels. To establish whether or not this is the case, in 2010 the G20 Summit Leaders asked the IEF, IEA, OPEC and IOSCO to produce a preliminary joint report on how the oil spot market prices in general are assessed by oil price reporting agencies (PRAs) and how this affects the transparency and functioning of oil markets. These four organisations appointed the author of this book, Liz Bossley, and Dr. John Gault of the Graduate Institute of Geneva to prepare a report considering this issue in 2011.

Because of the disproportionate influence of Brent indicated by the numbers above, the report focussed on this key benchmark grade. Consilience's work for the G20 remains subject to a confidentiality undertaking. However, after receiving the report IOSCO launched a consultation on 1st March 2012 concerning the "Functioning and Oversight of Oil Price Reporting Agencies". Consilience's response to that consultation is not confidential[24].

On 5th October 2012 IOSCO published its "Principles for Oil Price Reporting Agencies" with some motherhood and apple pie recommendations. These are reproduced in the insert box below.

---

**The PRA principles require:**

• The formal documentation and disclosure of all criteria and procedures that are used to develop an assessment, including guidelines that control the exercise of judgment, the exclusion of data as well as the procedures for reviewing a methodology. This information facilitates the evaluation of the impact of a methodology on the reliability of an oil derivatives contract. **[Methodology]**

• Transparency of procedures by which PRAs will advise stakeholders of any proposed changes to a methodology, including the opportunity for stakeholder comment on the impact of any changes. This is critical to allow a market authority to determine whether such changes, if proposed, may affect the reliability of a derivatives contract or otherwise result in possible market disruption. This principle also calls on PRAs to routinely re-examine methodologies to ensure

---

their continued reliability. **[Changes to a Methodology]**

• PRAs to give priority to concluded transactions in making assessments and implement measures intended to ensure that the transaction data submitted and considered in an assessment are *bona fide*, including measures to minimize selective reporting. These measures are intended to promote the quality and integrity of data and in turn the reliability of assessments. **[Market Data Used in Price Assessments]**

• Procedures to ensure the integrity of information, including procedures to set standards for who may submit data, quality control procedures to evaluate the identity of a submitter and internal controls to identify and respond to improper communication between submitters and assessors. These measures are intended to promote the accuracy and integrity of assessments. **[Integrity of the Reporting Process]**

• The adoption of guidelines to ensure the qualifications of assessors, including their training and experience levels. This principle also addresses continuity and succession planning in respect of assessors. These measures are intended to promote the integrity and consistency of assessment. **[Assessors]**

• The institution of internal controls requiring on-going supervision of assessors and procedures for internal sign-off on assessments as a means to promote the integrity and reliability of assessments. **[Supervision of Assessments]**

• The contemporaneous documentation and retention for five (5) years of all relevant information and judgments made in reaching a price assessment, including any exclusions of data. This is intended to facilitate inquiries by market authorities. **[Audit Trails]**

• The documentation, implementation and enforcement of measures to avoid conflicts of interest. These measures are intended to insulate assessments from improper influences, such as commercial or personal interests of the PRA or any of its personnel. The measures call for the functional and operational separation of a PRA's assessment business from any other business that may present a conflict of interest. These requirements are intended to protect the integrity of assessments. **[Conflicts of Interest]**

• A written and published procedure for receiving, investigating and retaining records concerning complaints about a PRA's assessment process, including recourse to an independent third party **appointed by the PRA**. [Emphasis added by Consilience]. The principle also requires that details concerning complaints should be documented and retained by a PRA. These measures are intended to promote the reliability of assessment methodologies through stakeholder input and alert a market authority to possible factors that might affect the reliability of assessments. **[Complaints]**

- A commitment to make available to relevant market authorities audit trails and other related documentation. This is intended to facilitate a market authority's ability to access data that are needed to determine the reliability of a given assessment referenced in an oil derivatives contract or to access information that might be needed to investigate and prosecute illegal conduct affecting a derivatives market. **[Cooperation with Regulatory Authorities]**

- An annual independent external auditing of a PRA's compliance with its methodology criteria and the requirements of these principles, which should be published. This is intended to encourage compliance with the principles and provide additional confirmation to market authorities of such compliance. **[External Auditing]**

In the light of recent reporting issues with key benchmark commercial data, e.g. LIBOR, it may not be current best practice to have the valuation of such a vast quantity of oil dependent on any one price benchmark, particularly in the absence of a standing external oversight body. Irrespective of the bona fides of the PRAs collecting and reporting the data, the fact remains that the data set on which those PRAs can call is very small: on many days no actual deals are transacted to inform the process. Very few market actors supply data to the benchmark price establishment procedure.

IOSCO pointed out in its 5th October 2012 report, referred to above, that it has a brief to consider the derivatives market, not the physical market. The prices that are taken up and used as a benchmark to set the value of the absolute price, A, in contracts around the world are physical/OTC forward prices. These prices are then used by the futures and derivatives market. The physical market is unregulated.

## Determining the Value of A: Forwards and Futures

This Chapter is examining the absolute price of oil, A, and we have so far mentioned the benchmark grades that are used in setting the value of A. The rest of this Chapter will look at the instruments that are used to establish and to hedge the value of A; the forwards and futures markets. These are instruments that trade at a fixed and flat price and therefore have a role to play in solving the price formulae contained in physical contracts.

# The Forward Market and the Value of A

A forward contract is an OTC commitment between a seller and a buyer to deliver and take delivery of a cargo of a specified grade of oil during the course of a specified forward period, usually of a month, at a fixed price. The month in question can be 2, 3 or more months forward from the transaction date. In the OTC market there is no limit to how far forward counterparties may deal; the only limit is how far forward a trader can find a counterparty that is willing to take the other side of the transaction.

Dealing in a forward contract by definition means that the cargo that is eventually delivered cannot be identified at the time the deal is struck. This is because the relevant lifting terminal operator does not allocate individual cargoes to individual producers until a time during the month prior to the month of loading of the cargo that is the subject of the contract. The precise date of allocation and the allocation procedures vary from country to country and from terminal to terminal, as discussed above.

For a forward contract to be successful there have to be many buyers and many sellers all trading standardised cargo quantities under standardised general terms and conditions of sale. If these conditions are met then market participants can organise their physical supply and manage their forward price risk safe in the knowledge that if they open a position by dealing in the forward market, they will be able to close it again at a time of their own choosing without being held to ransom by a dominant player and without incurring legal basis risk, i.e. they can buy and sell on back-to-back contractual terms.

In the crude oil market OTC forward contracts emerged before there were regulated futures exchanges offering oil price risk management solutions. The two most successful forward contracts for crude oil have been those for Dubai and Brent. As Dubai production has declined to less than 100,000b/d the Dubai forward contract has ceased to trade and has been eclipsed by a partials market, as described above in the discussion of the Dubai benchmark, which can be cash-settled or used to accumulate a deliverable quantity of oil. The forward contract in "Brent" still exists, but is not as active as it was at its peak in the mid-1980s. Nevertheless, despite its shortcomings, it remains a key component of the Brent benchmark discussed in the previous section of this Chapter. Today it is known as the 25-Day BFOE contract.

**The 25-Day BFOE** contract is a forward contract for 600,000 bbl cargoes of either Brent, Forties, Oseberg or Ekofisk with a three day loading date range that may be from at least 25 days in the future up to about 6 months in the future. The actual three day loading date range and the grade of the cargo to be delivered by the seller is not known at the time of the transaction. These are only confirmed by the seller 25 days in advance of loading. It trades at a fixed and flat price and is a common reference point in establishing the value of the absolute price, A, in crude oil physical contracts.

The forward contract in Brent emerged in 1981. It now trades under the general terms and conditions issued by Shell UK – the SUKO 90 terms[25]. It was originally a contract whereby cargoes, then of 500,000 bbls +/- 5%, of Brent System crude oil[26] could be sold well in advance of the date when the terminal operator allocated cargoes to Brent producers in the monthly lifting schedule.

Sellers would sell a cargo for delivery in a specified month, say, January 1983 and were only obliged to let the buyer know which of the then many cargoes of Brent it would deliver *15 clear days before the delivery date*. Once the lifting schedule was issued, in those days around 15th M-1, or in this example, by 15th December 1982, the producer that had been allocated the cargo with a delivery date range of 1st-3rd January 1983, would tell the forward contract buyer that it was the "1-3 January" cargo of Brent it would be receiving. Thereafter the accomplishment of the contract proceeded in accordance with the physical ship nomination and cargo documentation procedure described in Chapter Two. On 16th December 1982 the producer that had been allocated the cargo dated 2nd-4th January 1983, would tell the forward contract buyer that it was the 2-4 January cargo it would be receiving and so on until the 29-31 January cargo had been passed on to a 15-Day Brent cargo buyer.

## Cui Bono? Who Benefits from Trading?

A popular misconception about forward contracts is that each cargo is traded many times over with each trader attempting to take a margin as the cargo is passed on down a chain from a producer to an eventual refiner. In fact a chain only forms at the 11th hour when *producers* take the decision to fulfil a forward contractual commitment to sell a forward cargo that it may have taken many months earlier. This is referred to as a producer "wetting a chain". The physical cargo passes from the producer and on from trader to trader as this wetting process unfolds. Each company that receives the cargo may have entered into the transaction at a different period of time in the past and for very different reasons.

---

[25] "Agreement for the Sale of Brent Blend Crude Oil on 15 day terms Part 2 General Conditions Shell U.K. Limited July 1990"

[26] The term Brent Blend was only coined in August 1990 when Brent System crude oil and Ninian System crude oil were commingled in storage at Sullom Voe. Before that date the two streams were sold separately.

For example, say BP wants to wet a chain in the April forward contract, i.e. to supply a 15-Day Brent contract with one of its own cargoes derived from its equity production. It may decide to pass that April forward cargo to, say, Vitol. BP and Vitol may have entered into that forward contract for delivery of an April forward cargo back in November of the previous year. Vitol may have bought an April forward contract back in November to hedge its expected purchase of a physical Caspian cargo for April delivery oil under a term contract. Let's say the price of the April delivery forward contract between BP and Vitol was $80/bbl in November of the previous year.

Now Vitol has the April forward contract from BP it may decide to use it to supply a contractual commitment to supply an April delivery forward contract to, say, Mercuria. Let's say Vitol made the sale of the April delivery forward contract to Mercuria in January at a price of, say, $95/bbl. Vitol may have made that sale of the April delivery forward contract in January to hedge its sale of perhaps an Angolan cargo to a Mediterranean refiner. Mercuria may have purchased the April delivery forward contract in January to hedge its purchase of a North Sea cargo that it planned to take to the US Gulf as part of a transatlantic arbitrage.

Once the chain has formed, a process explained in the next sub-section, an outside observer might be tempted to say that Vitol has made a killing of $15/bbl in the April delivery forward contract by buying a cargo at $80/bbl in November and selling it at $95/bbl in January. Nothing could be further from the truth. Whether or not Vitol is making a profit or loss will depend on how its transactions on the Caspian cargo and the Angolan cargo played out.

When Vitol passed on the April delivery forward cargo supplied to it by BP, it might instead have passed it on to another trader or oil company to whom it had sold an April delivery forward contract. For example, Vitol may have sold an April delivery forward contract to Shell in December at a price of $70/bbl to hedge its sale of a Gulf of Suez cargo to an Eastern European refiner. It may choose to use the April delivery forward cargo supplied to it by BP to fulfil its commitment to deliver an April delivery forward contract to Shell.

In this case an outside observer might conclude that Vitol had made a thumping great loss of $10/bbl by buying a cargo at $80/bbl in November and selling it at $70/bbl in December. Whether or not Vitol is making a profit or loss will depend on how its transactions on the Caspian cargo and the Gulf of Suez cargo played out.

All that can be said for certain from the outside looking in is that Vitol had used the April delivery forward contract to manage its absolute price risk, A, in a number of transactions. How it managed its time differential risk, T, and its grade differential risk, G, and how much margin it locked in is not apparent from looking at how it balanced its many physical purchase and sales commitments in the April delivery forward market. So to those who believe that "traders are eating their lunch" I would suggest that the traders are not only sitting at a different table, they are probably dining in an entirely different restaurant.

## Forming Brent Chains

The chain formation process is only about the balancing of all the purchase and sales commitments entered into by all actors in the April delivery forward contract in the six or so months during which the contract traded. No inference can be drawn about which companies are making profits and which are making losses from examining the chain after it has formed.

In its heyday in the mid-1980s, the 15-Day Brent market was trading 40 or more cargoes per day, i.e. many more cargoes of Brent than were actually produced. So when a producer passed on the dates of a cargo to a buyer that had previously bought a forward 15-Day Brent cargo from it, that cargo was then passed on further to a subsequent buyer to whom the first buyer had sold. This was then passed on to the second buyer's subsequent buyer etc. Each cargo could end up being passed down great long chains of 50 or more companies before a refining buyer of the 15-Day Brent cargo would decide that the specific Brent cargo delivered on those particular dates suited its refining system and would take delivery of it.

The decision by a refiner to take delivery of a 15-Day Brent cargo and not pass it on before the 15 day notice period expired at 5pm London time on the 15th day before delivery, effectively took that cargo out of the 15-Day market and converted the cargo instantaneously to a Dated, physical or wet cargo. A Dated, physical or wet cargo was in those days a cargo loading *within* 15 days.

Each forward contract chain must start with a producer and end with a company taking delivery of the physical cargo onto a ship. Any other companies getting involved in the market must ultimately buy and sell an equal number of cargoes before the contract expires to be balanced, otherwise they will find themselves in breach of contract on one side or the other.

As time moved on to the late 1980s declining Brent system crude oil production left the 15-Day market open to "squeezes", i.e. traders buying up a lot of forward 15-Day cargoes and retaining cargoes as they were passed down chains to them. This meant that companies seeking to close "short" 15-Day Brent market positions could not find companies willing to supply them with cargoes that still qualified as 15-Day Brent. They would have to pay a substantial premium to the "long" 15-Day Brent trader that was squeezing the market. This terminology may be confusing to some, so in simpler terms here's how it worked.

Any trader, say Company X, that had the malicious intention of squeezing the market would over a period of months or even years buy up a lot of 15-Day Brent cargoes for delivery in a specific month, M. That trader X is now "long" of 15-Day Brent cargoes for delivery in month M.

When it got to 15th M-1, the dates of Brent cargoes for loading in month M would be issued to Brent producers by the terminal operator. Each producer could now start to declare its intention, if it so wished[27], to supply its own cargo available for lifting at Sullom Voe in a specified three day date range in month M to a company that had previously bought a 15-Day Brent cargo from it. In other words physical cargoes of Brent would start to pass down 15-Day Brent chains in time to meet the date declaration deadline of 5pm London time 15 clear days before the delivery date.

Each time one of these three day date range cargo declarations came in to trader X it would not pass the cargo on to a 15-Day Brent buyer, but would take the cargo out of the chain and sell it, probably the next day, as a physical Dated cargo that no longer qualified for supply into the 15-Day market.

Any trader that had sold the 15-Day Brent contract for delivery in month M to trader X in previous months would be "short" of the 15-Day Brent contract for delivery in month M to trader X. As fewer and fewer cargoes became eligible for supply into the 15-Day market over the passage of time any company, call them company Y, that was short to trader X, began to find it increasingly difficult to find anyone in the market who would sell a cargo of 15-Day Brent to it that it could use to fulfil its commitment to supply trader X with a 15-Day cargo. Ultimately the only company in the market that had 15-Day Brent cargoes for delivery in month M to sell would be trader X themselves. Trader Y would have to

---

[27] Brent producers had no obligation to use the 15-Day Brent market, nor have producers of Brent, Forties, Oseberg and Ekofisk any obligation to supply cargoes into the 25-day BFOE market today. But if a producer has a cargo of one of these grades that still qualifies for declaration into the 25-Day BFOE market it may choose to sell its cargo as a 25-Day BFOE cargo rather than a dated cargo if this gives it a higher price.

pay an inflated price to trader X to close its short position. This is a classic squeeze. It was to prevent squeezes of this nature recurring that the Brent base of the forward market needed shoring up, as discussed above, and may need shoring up again in the near future.

This behaviour resulted in two consequences:

- as trader X took cargoes out of 15-Day Brent chains and dumped them into the Dated Brent market, the price differential between Dated Brent and 15-Day Brent widened, as Dated Brent moved down and 15-Day Brent was squeezed up. In other words the market began to exhibit a steep contango formation that was not justified by supply and demand fundamentals, but only came about by the artificial squeeze on the 15-Day Brent contract. This was the trigger for the introduction of the "dated-to-paper" CFD market that allowed companies to hedge their Dated Brent versus 15-Day Brent time differential price risk, T. This development will be discussed in the next Chapter of this book;

- market participants looked for ways to beef up the volume of cargoes that qualified for delivery into the 15-Day Brent market to make it more difficult and costly for companies like trader X to squeeze the market. The more Brent cargoes there were, the greater the number of 15-Day cargoes that trader X would have to buy and the greater the offsetting loss it would make when it had to dump into the Dated market those excess cargoes that it was taking out of 15-Day chains.

## 15-Day Brent and the Metamorphosis into 25-Day BFOE and Beyond

The 15-Day Brent contract has evolved over time to shore up declining Brent volumes and to try to make the contract unsqueezable:

- In August 1990 the Ninian system crude oil stream was commingled into the Brent system crude oil stream forming a much larger volume of "Brent Blend". This would have happened anyway because it made considerable sense for upstream producers of Brent and Ninian to rationalise Sullom Voe operations to cut costs;

- In 2002 Forties Blend and Oseberg Blend were included as substitute grades in the assessment of the value of the benchmark Dated Brent price and as deliverable grades in the Brent forward contract. To help refiners cope with the fact that, although they had bought Brent they may end up receiving Forties or Oseberg and may have to trade out of these grades and into Brent in the Dated market, an additional 6 days' notice of the cargo date range and grade to be delivered was introduced. So what had been 15-Day Brent became 21-Day Brent/Forties/Oseberg or "21- Day BFO";

- In 2007, Ekofisk was added to the range of grades of crude oil that were deliverable in the 21-Day contract, making it "21-Day BFOE";

- In 2012 the notice period for the dates and the grade to be delivered into a forward contract was extended from 21 days to 25 days, making it the "25-Day BFOE";

- Platts has signalled its intention from 2015 to report only those forward contracts that provide 30 days' notice;

- On 8th February 2013 Shell introduced a series of price escalators to encourage the delivery of grades other than the lowest-priced Forties into the 25-Day BFOE contract. This is discussed in more detail in Appendix 1.

During the course of this evolutionary process there were some other changes agreed to the Brent forward contract. The cargo size was increased from 500,000 bbls to 600,000 bbls but the operational tolerance was reduced to +/-1%. The notification deadline for declaring the three day date range and grade of a cargo to be delivered was advanced from 5pm London time to 4pm London time at the request of European traders that generally operate one hour ahead of London.

As mentioned above, Platts is considering using a wider range of international grades of more disparate quality in setting the price assessment of physical Dated Brent and as a substitute for Brent in the forward, 25-Day BFOE contract. As discussed above, in Consilience's opinion any such concerted change by the industry would benefit from some regulatory over-sight to ensure that all stakeholders have a voice in any changes.

## Credit Security in the Forward market

The contract size in forward contracts tends to be large and indivisible. As mentioned above the 25-Day BFOE contract trades in cargo sizes of 600,000 bbls with a +/- 1% operational tolerance on the volume. At a price of $100/bbl this implies a transaction value of $60 million.

Unsurprisingly companies are unwilling to extend that level of credit to just any old Tom, Dick or Harry. Typically the finance director in any given company has the ultimate responsibility for approving the credit limit for any of the company's likely counterparties. This can be very frustrating for the traders on the desk who want to deal with the counterparty that gives them the best price. Good terms ultimately mean good bonuses for the trader. It is doubly frustrating when the company has a large credit limit with a counterparty, particularly with the banks that trade in the oil market, but it has been used up by treasury colleagues for an interest rate or FOREX deal.

Obviously there is no point in getting great terms on a deal if the counterparty does not pay up. So in many companies cargoes will only be sold to companies that will provide a letter of credit (LC) to guarantee payment. Oil companies will demand LCs from private trading houses that appear to be much bigger and more successful than the oil companies selling the oil. The difference is that trading companies, with some notable exceptions, do not publish an annual report and accounts so it is impossible to judge the company's creditworthiness with any certainty. So the requirement for the buyer to open an LC is an essential condition of a sales contract.

If an LC is required, it has to be opened by a first class international bank acceptable to the seller. The seller will be calling on the bank for payment so if the bank is not creditworthy the LC is valueless.

The LC has to be in place at least 10 days before the cargo is due to load. If it is not, the seller may wish to suspend delivery and sue the buyer for breach of contract rather than risk loading the oil onto the buyer's vessel without a payment guarantee being put in place. If the seller wishes to suspend the contract it needs to have time to put another sales contract in place and for the new buyer to nominate a ship. Otherwise the seller will be in the position of having to make a distressed sale at a low price in order to avoid a failure to lift.

There are two basic types of LC:

- An Irrevocable Documentary Letter of Credit; or,

- A Stand-by Letter of Credit.

The Irrevocable Documentary LC is designed to protect the seller so once it is put in place the buyer cannot alter its terms without the seller's consent: hence the term "irrevocable". Once the LC is in place responsibility for payment rests with the bank, not the buyer.

The seller presents cargo documents and an invoice direct to the bank to achieve payment. If the cargo documents are presented in accordance with the contract and complying with the buyer's documentary instructions, the bank is obliged to pay, even if the buyer has gone bust in the 30 days between delivery of the cargo and the payment due date: hence the term "documentary". It is crucially important that the cargo documents are accurate as to the last detail. If they are not the bank may attempt to use any loophole to avoid the obligation to make payment. For this reason, sellers claiming payment under an LC often produce a Letter of Indemnity (LOI), rather than the actual cargo documents in order to effect payment. We will come to LOIs shortly.

The Stand-by LC is also designed to protect the seller so again must be opened with a first class bank acceptable to the seller. In this case the seller presents the documents and the invoice to the buyer and only if the buyer does not pay up does the seller call on the Stand-by LC. In this case the cargo documents have already been sent to the buyer so the bank pays up against an LOI and an invoice.

It is not unusual for original cargo documents to be delayed in-transit and for them not to be available for presentation to the bank or the buyer by the due date for payment. The terms of the LC provide for this common situation by allowing the seller to produce an LOI warranting that it has good title to the oil it has delivered and pledging to hold the buyer harmless against any third party alleging a claim on the oil.

Typically sellers present an LOI to receive payment, even when the cargo documents are available. It is easy for the seller to change an LOI if it contains an error: it is much harder to get the loading terminal to change original documents that have to be signed by the master of the vessel if these contain an error. Once payment is safely received the original cargo documents are then passed on.

In the absence of cargo documents an LOI is usually acceptable. If the seller is not a credit worthy party, the buyer will require that the LOI be endorsed by a first class bank acceptable to the buyer. If there is a claim against the oil there is no point in calling on the seller for legal protection if the seller has gone bust.

LCs have to be monitored closely. Buyers will typically try to design the LC to cover the minimum necessary quantity of oil deliverable under the contract and for the LC to expire as soon as possible after the delivery date. Sellers have to ensure that the LC covers the maximum quantity of oil that the master of the vessel might choose to load for safety reasons and to ensure that the LC does not expire before the payment date, even if the cargo loading is deferred for any reason.

The cost of LCs is extremely variable depending on the credit standing of the buyer concerned with the bank providing the financial security, but the order of magnitude is 2-5 cents/bbl.

We will now turn to the other main category of contract that is used to set the value of the absolute price, A, in physical contracts – the futures markets.

## The Futures Market and the Value of A

The two futures markets of most relevance to the oil industry are the Intercontinental Exchange's (ICE) Brent futures market[28] and the Chicago Mercantile Exchange (CME)/ New York Mercantile Exchange's (NYMEX)[29] light, sweet crude oil markets. The latter is often referred to as the "WTI" contract, as explained earlier in this Chapter when we discussed the WTI benchmark. There is a third relevant futures contract in the crude oil market, namely the Dubai Mercantile Exchange's (DME) Oman futures contract. The Dubai Mercantile Exchange (DME) is the newest of the three and was launched in June 2007, as a joint venture between the Dubai and Omani governments and the CME.

The contracts of these three exchanges are used by crude oil traders as the benchmark price to establish the absolute price, A, under a physical sales contract (by referring to the exchange's price in the cargo sales price formula), for hedging a range of physical grades that use these benchmarks and for speculative purposes.

The ICE and CME futures contracts are regulated by the UK Financial Services Authority (FSA) and the US Commodity Futures Trading Commission (CFTC). Both exchanges offer very similar futures contracts in Brent, but the ICE contract is dominant in the Brent market. ICE also offers a US light, sweet contract, but CME is dominant in WTI. The DME is regulated by the Dubai Financial Services Authority (DFSA) and it is a much less liquid contract trading a fraction of the volume of either of the larger competitors[30].

The ICE Brent contract trades consecutive months out to the December 2016 contract. Thereafter the June and December contract months are listed up to and including 2019. The later contracts trade but are less liquid than the earlier monthly contracts. The ICE

---

[28] Previously known as the International Petroleum Exchange ("IPE"). ICE bought IPE in May 2001.

[29] The Chicago Mercantile Exchange bought the New York Mercantile Exchange in August 2008.

[30] According to DME, both the Dubai and Oman governments price their physical sales going to the Far East by reference to the DME Oman contract. The DME is lobbying hard to have other OPEC governments reference their Far Eastern sales to the DME Oman contract rather than the Platts Dubai quotes. This may change the fortunes of the exchange in the future but in the meantime the contract is small in comparison with the ICE Brent or CME NYMEX WTI crude oil contracts trading under 10 millionb/d.

Brent New Expiry (Brent NX) contract trades consecutive months out to a maximum of 72 months forward, with a new year being added on the expiry of the December contract at the front-end each year. Thereafter the June and December contract months are listed for an additional two calendar years.

This Brent NX contract was introduced when Platts changed the expiry date of the 21-Day BFOE contract to 25-Day BFOE. The Brent futures contract cash-settles by reference to the forward contract. In time it is planned that these two contracts - Brent futures and Brent NX futures - will converge with the Brent NX timetable applying after the final Brent contract has expired. From March 2015 and beyond the expiry date of the Brent NX contract will be the last business day of month M-2 for month M. This will accommodate an anticipated shift from 25-Day BFOE to 30-Day BFOE in the forward market from that date.

The CME WTI contract trades nine years forward arranged as consecutive monthly contracts in the current year and the next five years. Thereafter the June and December contract months are listed beyond the sixth year.

The DME Oman contract lists monthly contracts in the current year and the next five calendar years. A new calendar year is added following the expiry of the December contract of the current year.

A futures contract is similar to a forward contract in that two parties agree to pay a certain price for oil for delivery in a specified time period in the future. To facilitate trading, the exchange determines standardised specifications for the contract and guarantees payment and performance. The nature of this guarantee is that the exchange, or more correctly the clearinghouse, becomes the central counterparty to every transaction. In effect the exchange does back-to-back deals for every contract for oil that is bought and sold.

So if company X sells 50 lots (NB: 1 lot=1,000 bbls) of Brent onto the ICE market, these 50 lots might actually be bought by company Y. But company X actually sells to the exchange and company Y simultaneously buys from the exchange at the same price. Neither party knows who is taking the other side of the deal. In the case of ICE the clearinghouse is ICE Clear Europe. In the case of both CME and DME contracts are cleared and settled by CME Clearing.

This might suggest that when a party buys or sells a futures contract, the clearinghouse is exposed to the creditworthiness and performance of that company. Clearinghouses have very robust procedures in place to ensure that they do not make a loss if a company trading in its contract defaults on performance or payment. Usually futures trades undertaken by a company must be registered and held by:

- a dealer participant, authorised to deal on their own behalf or on behalf of clients; or,

- a clearing member of the exchange, i.e. a company that has rights to a share of clearing fees, but which also takes risk in providing a share of the guarantee fund that under-writes the financial security of the clearinghouse (See Figure 19).

**Figure 19 Dealing with the Clearinghouse**

Client    Dealing Participant    Client
          Clearing Member

These entities charge the client brokerage and/or clearing fees. The actual flow of money is to or from the client through the dealing participant/clearing member from or to the clearinghouse. Unless the company is a dealing participant/clearing member itself it does not have a direct contract with the exchange or the clearinghouse. The dealing participant/ clearing member is exposed to credit and performance of its client counterparty in implementing its orders on the exchange under the terms of a services contract between the client and the dealing participant/ clearing member. The client counterparty is equally exposed to the dealing participant/ clearing member.

The regulator requires companies offering such services to be properly authorised and, amongst other things to hold clients' funds in segregated accounts. The intention is that, if the brokerage/clearing service provider goes bust, the clients' funds are protected. If it does not protect client funds in this way a defaulting brokerage/clearing service provider is liable to criminal prosecution and civil penalties imposed by the regulator.

# Lessons from MF Global

This protection proved inadequate for market participants dealing through MF Global Holdings Ltd. When the brokerage firm went bust in October 2011 it was discovered that $1.6 billion of client funds could not be located. This matter is subject to on-going civil and criminal investigation. In the meantime clients whose money disappeared had to finance the sudden cash shortfall themselves. Some went bust in the short-term, deriving little comfort from the fact that some of their money may be retrieved by the administrators in the long-term. Users of futures exchanges should bear the MF Global case in mind when they are told that futures exchanges provide financial and performance guarantees.

## Initial Margin

When a company opens a new position on a futures exchange through a dealing participant or a clearing member, regardless of whether it is a long or short position (i.e. buying or selling), it must post a good faith dealing deposit known as its "Initial Margin" with the clearinghouse via its dealing participant/ clearing member.

The size of this initial margin varies with the size of the open position (i.e. it applies to each contract, so more contracts means more initial margin) and with the actual volatility of the contract in question. Volatile contracts expose the trader to more risk so the exchange demands a higher initial margin. The calculation of the riskiness of a futures contract, which informs the setting of initial margin, is done using Standard Portfolio Analysis of Risk (SPAN) software. The size of the initial margin required by the clearinghouse can be changed at any time the exchange deems that the contract has become more risky.

If the company enters into a "spread trade", i.e. it buys one contract month and sells a different contract month, a lower level of initial margin is required. This is because it is likely that when a company is long of a contract in one month and short of the same contract in another month, if the market moves any losses the company makes on one trade will be offset by gains on the other trade.

Initial margin is returned to the trader when it closes its position. If the trader defaults on a contract the initial margin is retained by the clearinghouse to offset its losses.

## Variation Margin

Each company's open positions are "marked to market" every night. For example, if a company sells a contract at $100/bbl and the closing price that night is $100.50/bbl, the position is making a notional loss of $0.50/bbl. Cash to cover this notional loss must be lodged with the clearing broker next day. This is called posting "Variation Margin". If the following night the price closes at $99.50/bbl, the position, opened at $100/bbl, is now making a $0.50/bbl profit and the Variation Margin repaid to the principal is $1/bbl, i.e. the repayment of the $0.50/bbl loss from the previous night plus the additional $0.50/bbl profit.

If a principal does not pay its initial or variation margin promptly, this may be a strong signal that the company may be in financial trouble and the exchange, or the broker, can close the position before losses escalate out of control.

Typically a back office member of staff is responsible for monitoring margin calls and ensuring that margins are paid promptly in order that the exchange does not close any positions for want of a margin payment.

Only companies with liquid cash availability should deal on the regulated exchanges. Examples are legion of companies that were making money overall on a combined physical plus hedge position, but went bust because they could not meet daily margin calls on their hedge position in the futures market. For this reason the futures market may not be an ideal price management tool for companies that do not have substantial cash liquidity.

Figure 20 shows the flow of information in exchange-based dealing.

**Figure 20 Information Flow for Exchange-based Dealing**

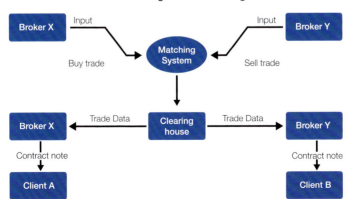

## Physical Delivery

**The ICE Brent futures contract** deals in lots of 1,000 bbls, therefore delivery cannot take place because 1,000 bbls is too small to ship. All transactions are settled in cash by entering into an equal and opposite position, i.e. if a principal buys 1 lot of ICE Brent, it must subsequently sell 1 lot of ICE Brent before that contract expires. Any position that is not closed when the contract expires is cash-settled at a price determined by the exchange, based on market data for the 25-Day BFOE market on the date of expiry.

Trading for any given month on the ICE Brent contract ceases at the close of business on the business day immediately preceding the tenth day prior to the first day of the delivery month M if such tenth day is a banking day in London, i.e. on 10th M-1. In other words expiry of the April contract is on or around 10th March, depending on weekends and bank holidays. If the tenth day is a non-banking day in London (including Saturday), trading ceases on the business day immediately preceding the next business day prior to the tenth day.

The precise formula for determining the cash settlement price is a secret closely guarded by the ICE, although it is typically not difficult to work out. This is to avoid any attempt by traders with large open positions to manipulate in their own favour any of the data sources used by the ICE to arrive at a fair cash settlement price that reflects the value of the underlying commodity on which the ICE Brent contract is based, i.e. 25-Day BFOE.

When the 21-Day BFOE contract was changed by Platts to 25-Day BFOE in January 2012 a new contract, Brent New Expiry (Brent NX), was introduced to reflect the new timetable. Trading in the NX contract ceases at the close of business on the business day immediately preceding the twenty fifth calendar day preceding the first day of the contract month, if such twenty fifth day is a Business Day. If the twenty fifth day is a non-banking day in London (including Saturday), trading ceases on the business day immediately preceding the next business day prior to the twenty fifth day. As mentioned above the Brent NX expiry date will move to the last business day of month M-2 for month M from March 2015.

## EFPs and Physical Delivery

Although it is true that a cargo cannot be delivered to settle open positions in the ICE Brent contract, ICE argues that its Exchange for Physical (EFP) mechanism is de facto physical delivery. An EFP represents a physical deal between two named counterparties transacted off-exchange. The counterparties may agree to transfer the physical position onto the exchange by entering into an equal and opposite position in the physical market and at the same time re-opening the original position in the futures market. The price at which the position is transferred does not feature in ICE price information shown to other market participants because its relevance has not been tested in the wider market. However the volume is recorded and the price is used to calculate the subsequent variation margin payments of the two counterparties to the EFP. Each party to the EFP can now close its futures position at a time of its own choosing without consultation with the other party to the EFP.

**The CME NYMEX US light, sweet futures contract** operates in exactly the same way with one key difference: 1,000 barrels of oil is deliverable in the US pipeline system, so the CME Nymex contract can be settled by making physical delivery. As discussed above WTI is a long-standing US benchmark grade of crude oil and is the deliverable grade in the CME Nymex Light Sweet Crude Oil Contract. Trading in the contract for delivery month M ceases on the third business day before the twenty-fifth M-1. Hence for the contract month of April the contract would expire on the 22$^{nd}$ March, subject to weekends and bank holidays.

The CME light, sweet contract can be cash settled by closing positions before expiry or can be settled by making/accepting physical delivery of the requisite quality of oil at Cushing. The deliverable quality is oil that meets the specifications of the TEPPCO pipeline and the Equilon pipeline, including foreign grades of oil that enter the US pipeline system at the Gulf Coast. This quality specification is API gravity between 37$^{\circ}$ and 42$^{\circ}$ and a maximum sulphur content of 0.42%. Hence what is still referred to in shorthand as "WTI" can in reality be crude oil from a number of different sources and countries.

The timetable for delivery is constructed to coincide with pipeline delivery scheduling in the US domestic market. It is the contract seller's (the short's) responsibility to ensure that the oil flows rateably over the month of delivery. The buyer (the long) has the right to call for delivery:

- By pump over into a designated pipeline or storage facility with access to seller's incoming pipeline or storage facility; or,

- By in-tank transfer of title to the buyer without physical movement of product, if the facility used by the seller allows such transfer; or,

- By in-line transfer or book-out, i.e. cash-settlement[31], if the seller agrees to such transfer.

The seller is required to give the buyer a pipeline ticket, any other certificates of quantity and all appropriate documents upon receipt of payment.

**The DME Oman futures contract** also provides for physical delivery, in this case of Oman crude oil FOB the Mina Al Fahal Terminal in Oman. Trading in the contract for any given month ceases on the last trading day of M-2. Hence trading in the April contract expires on the last trading day of February. A clearing party with an open short position that intends to make physical delivery must notify the exchange of its intention to do so one hour before the contract expires. However the clearing member is in contact with its customers in the days running up to expiry to organise this process and make sure that its customers actually do want to go to physical delivery rather than trade out of its position. A party with an open long position that intends to take physical delivery must notify its intention to do so by 14:00hours (New York time) on the first business day after the date of expiry on which the clearinghouse is open.

The clearinghouse matches the aggregate long and short positions of end-buyers and end-sellers into deliverable quantities. The volume that actually goes to physical delivery can be substantial in the range of 10-15 million barrels and the exchange nets off the biggest positions first. Positions are matched in parcels of at least 200,000 bbls +/- 0.2%, which may include exchange traded barrels or barrels sourced in the OTC market.

A physical supply contract is then deemed to have arisen between the end-buyer and the end-seller identifying the laydays (delivery window) of the physical oil concerned. Actual physical delivery then proceeds as normal for a physical oil contract and the terminal loading procedures, with the end-buyer nominating the tanker onto which the oil is loaded. This tanker is likely to have a capacity of considerably greater than 200,000 bbls, but it is up to the buyer to ensure that it acquires sufficient oil from other sources to ensure that it does not incur deadfreight costs[32] on the shipment. The Oman terminal sitting outside the Straits of Hormuz is frequently used to supply top-up parcels of crude oil onto tankers exiting less than fully laden from the Arab Gulf. The physical delivery mechanism of the DME contract is a regular source of such top-up parcels.

Unlike typical physical contracts, payment continues to be guaranteed by the clearinghouse. The physical buyer must post margin equivalent to the full value of the

---

[31] For full explanation of "book-out" please consult the glossary."

[32] Unused cargo carrying capacity on a vessel which must be paid for whether it is used or not.

cargo calculated as the volume multiplied by the price at which the buyer entered into the futures contracts, less any variation margin already lodged with the clearinghouse. This may be in the form of a letter of credit or other acceptable payment guarantee. The buyer is required to pay the seller for the oil within 30 days of the B/L date. If it does not pay on the due date the clearinghouse pays the seller using the buyer's margin or payment guarantee posted with it before physical delivery took place. The physical seller must post margin in a quantity determined by the clearinghouse, often about 10% of the value of the oil, sufficient to cover the costs of the seller failing to deliver. When actual delivery takes place, as evidenced by the bill of lading (B/L), the seller's margin is returned to it.

The three exchanges mentioned above also trade option contracts. Options will be discussed in Chapter Six.

# Hedging the Value of A

Forward and futures contracts may in certain instances be used by oil producers to sell physical cargoes of oil and by refiners to acquire physical feedstock, as described above. A more common use of these instruments is for oil industry stakeholders to hedge the value of the absolute price, A, in physical contracts. Hedging is in theory a very simple concept, but there are practical pitfalls for the unwary as described below.

## Basic Hedging Theory

Assume a refiner knows that it will want to buy a cargo of Crude X for delivery in four months' time. There is no forward market in Crude X so it cannot simply buy it today. There is a forward market in Crude Y. This forward market is trading in Crude Y for delivery four months forward at, say, $80/bbl. The refiner would like to buy Crude X for delivery in four months' time at $80/bbl, but it cannot do so because there is no forward market in Crude X. So today it buys Crude Y for delivery in four months, even though Crude Y is unsuitable for its own refinery.

In three months' time the market in Crude X is at last trading and the refiner can now acquire the cargo of Crude X that its refinery needs for delivery next month. Let's say that in the three months, while it was waiting, the price has risen to $100/bbl. The refiner must pay $100/bbl for Crude X.

But the refiner has also bought a cargo of Crude Y for delivery next month and the refinery has no use for it. So the refiner sells a cargo of Crude Y for delivery next month at the then current price of $100/bbl.

The overall financial position of the refiner is that it has acquired its cargo of Crude X at a net price of $80/bbl, i.e. $100/bbl to buy the physical cargo of Crude X and a $20/bbl profit from buying Crude Y at $80/bbl and subsequently selling it at $100/bbl.

If over the three months the oil price had fallen to, say, $60/bbl then the refiner would still have achieved a net price of $80/bbl. In this case it would have bought Crude X at $60/bbl and made a loss of $20/bbl on the purchase and subsequent sale of Crude Y. The hedge has still worked because the objective was to buy at $80/bbl. The refiner may wish it had not hedged, but the hedge has worked.

Would that all hedges were this simple. Once this basic hedging theory is tested in practice, some complications begin to emerge.

## Practical Complications

As an illustrative example, let's say Producer X wishes to lock in the price of a cargo of "Generic Blend" crude oil today, say July, that it will have to sell in, say, December. But it will be unable to sell this December delivery cargo until at least the first ten days of November when the terminal operator issues the schedule of cargo liftings for December. Producer X cannot sell Generic Blend for delivery in December in July because there is no forward market in Generic Blend. However Producer X is worried that the price of oil may fall from its level in July of, say, $100/bbl before early November when it is able to sell the physical Generic Blend cargo.

Let's say that physical Generic Blend is typically sold in the Dated, wet market with a formula of, perhaps, the published values of Dated Brent on the 5 days around the B/L date of the cargo plus a fixed grade differential of G, i.e. "Dated Brent (2-1-2) +G". The term "2-1-2" refers to the 5 publication days that are used to calculate the price formula. This means the two days before the B/L date, the B/L date and two days after the B/L date.

Producer X knows that its absolute price risk, A, is connected to the value of the Brent benchmark. But it has no idea in July what the B/L date of the cargo is going to be in December so there is no way at that time of managing the time differential risk, T. But the value of A can be managed by using the 25-Day BFOE market or the Brent futures market.

**Figure 21 July Forward Oil Curve**

Forward Delivery Month

Let's say that in July the forward oil curve for 25-Day BFOE is in backwardation as shown in Figure 21. In July the value of oil for delivery in December is $98/bbl, two dollars below the July price of oil for prompt delivery in July.

So in July Producer X can hedge the value of the Generic Blend cargo for delivery in December by selling forward at $98/bbl.

In the first ten days of November when the terminal operator issues the schedule of cargo liftings for December, Producer X will be establishing the date range of its Generic Blend cargo for delivery in December with the terminal operator of Generic Blend. Once it has done so it can now sell that Generic Blend cargo at the November price for deliveries in December.

Let's say between July and November that the level of the absolute price A has fallen and that backwardation has been replaced by contango. In November the value of the 25-Day BFOE cargo for delivery in December is now $80/bbl. See Figure 22.

**Figure 22 November Forward Oil Curve**

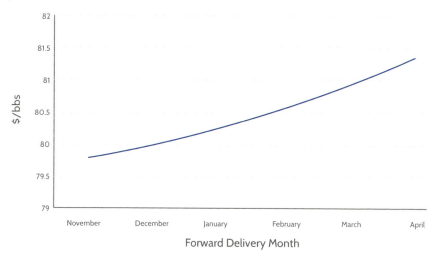

It is at this point that time basis risk in the hedge emerges, as explained below.

# Dated Brent Risk Hedged with a Forward Brent Contract

In November Producer X will be able to sell the Generic Blend cargo at a price that is loosely related to the 25-day BFOE price for December delivery of $80/bbl. The price formula under the physical Generic Blend cargo contract will be, as we said above, Dated Brent (2-1-2) +G. So, given the contango in the market, the price of Dated Brent, i.e. the price of Brent[33] for delivery in the next 10-25 days, will be at a discount to 25-Day BFOE for delivery in December. The earlier the dates of the Generic Blend cargo in December the lower the price Producer X will receive for it because of the contango value of T. Let's say that Producer X sells the December Generic Blend cargo at a price formula that delivers $79/bbl +G.

To complete the hedge Producer X must now take action. It has committed to deliver a cargo of 25-Day BFOE in December. It does not have a 25-Day BFOE cargo so it must buy one. The price of 25-Day BFOE cargo for delivery in December is currently, i.e. in November, $80/bbl. Producer X sold a December delivery 25-Day BFOE cargo back in July at $98/bbl and it is now covering that commitment by buying December delivery 25-Day BFOE at $80/bbl. It has an $18/bbl profit on the hedge to add to the sales price of the December delivery Generic Blend wet cargo at $79/bbl +G.

Producer X has an overall economic result of $79+18+G/bbl = $97/bbl +G. This is close to the economic result that Producer X wanted to receive when it sold the December

---

[33] Or more correctly the lowest of the prices of wet physical Brent, Forties, Oseberg and Ekofisk.

delivery 25-Day BFOE cargo back in July because it could not sell Generic Blend that far in advance.

If between July and November the absolute price, A, had risen instead of fallen, Producer X would be able to sell its Generic Blend cargo at a higher price, but would have to close its hedge by buying a 25-Day BFOE cargo for December delivery at a higher price too. Again its overall financial result would have been close to the $98/bbl +G level that it wanted to achieve back in July. It may wish that it had not hedged the cargo in these circumstances, but the hedge has delivered the price certainty Producer X sought when it opened the hedge.

## A Floating Priced Cargo Hedged with a Fixed and Flat Price Contract

The hedge described above is not a perfect hedge because the price of the physical cargo is linked to the price of Dated Brent, while the price of the hedge relates to a different Brent contract, the 25-Day BFOE contract. This gives Producer X "timing basis risk" in its hedge. We will demonstrate how this basis risk can be managed with a separate contract in the next Chapter of this book.

There is a further significant risk for Producer X associated with hedging the sale of a Generic Blend cargo with the sale of a 25-Day BFOE cargo. The price of the Generic Blend cargo is unlikely to be agreed as a fixed number for the whole cargo in November when the physical cargo Generic Blend cargo is sold by Producer X. We have said the the value of A in the price formula of the Generic Blend sales contract relates to the average price of Dated Brent as published on the 5 days around the B/L date, i.e. 2-1-2 pricing.

So if Producer X was to close the 25-Day BFOE hedge in November when it placed the Generic Blend cargo into a sales contract with a refiner, it would be exposed to the risk of the price of Generic Blend falling between November, when it bought a December delivery 25-Day BFOE to close its hedge, and the 2-1-2 dates around the loading date of the Generic Blend cargo in December. A perfect hedge requires first that *the pricing of the physical cargo* (not the execution of the physical sales contract) and the closing of the hedge take place as contemporaneously as possible.

Secondly, to perfect its hedge Producer X actually needs to buy the 25-Day BFOE cargo as an average of the prices published on on the 5 days around the B/L date of the Generic Blend cargo. This is not easy to do[34]. The 25-Day BFOE contract trades in single cargo lots of 600,000 bbls at the fixed and flat price at the time the deal is struck, not at a 5 day average price that would mirror the 2-1-2 price under the Generic Blend contract. So Producer X actually needs a hedge that will allow it to buy 120,000 bbls of 25-Day BFOE on each of the 5 days around the B/L date in December of the Generic Blend cargo. Such a contract does not exist.

This suggests that hedging with a futures contract might be a better solution for Producer X. This is because the futures contract trades in lots of 1,000 bbls as discussed earlier in this Chapter. Hence if Producer X wants to buy back 120,000 bbls of its hedge on each of the 5 days around the B/L date of the Generic Blend cargo it may be better advised not to hedge by selling a 600,000 bbls 25-Day BFOE contract, but to hedge by selling 600 lots of Brent futures and buying back 120 lots on the 5 days around the B/L date. This assumes that Producer X has the cash capacity to deal with futures contract margin calls, as discussed above.

## Which Month?

It should by now be apparent that selling a December delivery 25-Day BFOE contract is a flawed hedge of a physical contract that is also for delivery in December. This is because the formula price in a physical contract will establish the value of the absolute price, A, by reference to the forward contract month of January or even February, as it is published in December. By December the December delivery futures contract will already have expired. Trading in the December 25-Day BFOE contract will be very thin at the start of December and will have ceased altogether by about the 5th December.

Hence to hedge a December delivery physical cargo requires opening a hedge in a forward contract at least one month further forward (January or February) to ensure that the hedge can be closed at the same time as the physical cargo price formula is established by reference to prices published in December. If the December physical delivery dates of the Generic Blend cargo turn out to be late in December, the January futures contract will also have already expired and the 25-Day BFOE market in January cargoes will already be winding down. So a February 25-Day BFOE cargo or February futures contract would provide a more flexible hedge of a December physical cargo.

However, as indicated above, if the physical contract is priced by reference to the

---

[34] Most things can be done at a price. But buying a 25-Day BFOE cargo at an average price rather than a fixed and flat price would take the hedger into the partials market, which operates somewhat differently from the cargo market. An example of the partials market was discussed earlier in this Chapter in the description of the Dubai benchmark.

published values of Dated Brent, i.e. oil for loading in the next 10-25 days, there is a substantial risk that a hedge priced by reference to February delivery oil will not move cent for cent with the Dated Brent price in the physical contract. This concern is addressed in Chapter Four of this book. But for the moment suffice to say that when a company is hedging it needs:

- A high positive correlation between the price of the physical cargo and the price of the hedge instrument; and,

- A high negative correlation in the physical revenue stream and the hedge revenue stream.

Correlation indicates the strength and direction of a relationship between two variables, in this case the prices of different crude oil contracts.

- If two prices move completely in isolation from each other and the value of one has no impact on the other, then the correlation coefficient would be zero.

- If the two prices increase or decrease perfectly together so that a unit increase in the value of one price is reflected by a unit increase in the value of the other price, then the correlation coefficient would be +100%.

- In the case where a unit increase in the value of one price is reflected by a unit decrease in the value of the other price, then the correlation coefficient would be -100%.

## A Postscript on Speculation

We said earlier in this Chapter that a common use of forward and futures contracts is to allow companies to hedge the value of the absolute price, A, in physical contracts. These forward and futures contracts are also used by companies wishing to take a speculative position in the value of A.

After the banking crisis and the LIBOR-fixing scandal, the term "speculation" is often regarded as a dirty word. Consilience does not share this prejudice. Speculative involvement in the market is absolutely necessary in our opinion to provide the liquidity that the oil industry needs when it attempts to hedge. When a hedger enters the market to buy or sell oil as a hedge of a physical position it likes to see a range of competing counterparties prepared to take the other side of the deal. This means that the hedger can execute its transaction easily at a narrow bid-offer spread without having an undue influence on the price, even if it wishes to execute a large volume. If speculators were not there to provide liquidity it is questionable if oil contracts would have the narrow bid-offer spread of <$0.05/bbl that they currently enjoy or that they would be able to lay off large volumes of risk without moving the market.

In some press reporting the terms "speculation" and "manipulation" seem to be used almost synonymously. Obviously they are not. Backing a price view by opening a position and putting money at risk in the market is a very different thing from trying to misrepresent the true market value.

Others may feel that speculative involvement in the market is a bad thing because it exacerbates peaks and troughs in price trends. Do speculators make the oil market more volatile? Or is natural oil market volatility what attracts speculators to the market? This is a "chicken and egg" argument that is actually very difficult to resolve. It is to be hoped that in the regulatory drive to cut out market manipulation that speculation is not made more difficult.

## A Quick Recap

In this Chapter we have focussed on the value of the absolute oil price, A, as it appears in a physical oil contract. We have looked at the benchmark grades of oil that are used as reference points in physical contracts and noted their shortcomings. We have noted the declining volume under-pinning the three biggest benchmarks – Brent, WTI and Dubai - and the attempts to shore up these volumes by including other grades in the assessment of their price.

We also noted the logistical issues that have caused WTI to uncouple from the price of oil in the international seagoing market, which requires further infrastructural investment to cure. This investment may be forthcoming but whether or not the planning permission to build the necessary pipelines will be granted remains to be seen.

We have considered the particular case of Brent and the pervasive use of Brent-related instruments in the physical, forward, futures and derivative markets. We noted that big changes to this benchmark are envisaged, the first of which is explained in Appendix 1.

We have examined the two major contractual instruments that trade on a fixed and flat price basis. These are the forward and futures contracts, which allow floating price formulae under physical contracts to be resolved into a number expressed in terms of $/bbl.

Forward contracts are dealt OTC by named counterparties in an unregulated framework involving large contracts in cargo-sized lots. Futures contracts trade anonymously in lots of only 1,000 bbls on regulated exchanges, with financial and performance guarantees, of questionable value to companies that use the exchange, as the example of MF Global demonstrated all too clearly.

We have considered how forward and futures contracts can be used to hedge the value of the absolute price, A, in physical contracts and concluded that as a hedge of physical oil these contracts are something of a blunt instrument because they do not address the price differential risk represented by the value of T, i.e. timing risk.

Later on in Chapter Six we will look at additional contracts that involve dealing in the value of the absolute price, A, namely swaps and options. These additional contracts can be used by physical traders to hedge the value of A, but tend to be used by them to take out strategic large scale hedges, rather than to hedge the value of A in a single cargo with some exceptions. Swaps and options are also used by financial players to speculate in the value of A.

The next Chapter will consider the contracts that are used to set and hedge the value of the time differential, T, in the physical contract price formula.

# CHAPTER 4

❝❝ They say that time changes things, but you actually have to change them yourself ❞❞

*Andy Warhol*

---

As discussed in Chapter One the price of a cargo of oil is comprised of three components, the absolute price, A, the time differential, T, and a grade differential, G. This Chapter will concentrate on the value of the time differential, T, which represents the difference in value at a single point in time of the same oil if it were to be delivered in a variety of different future time periods. It is represented by the slope of the forward oil curve.

It will be recalled from Chapter One that at one instant in time when the price of oil for prompt delivery is at a premium to the price of oil for delivery at a later period, this is a formation referred to as backwardation. When the price of oil for prompt delivery is at a discount to the price of oil for delivery at a later period this is a formation referred to as contango.

Before isolating the value of T for further examination a word of explanation on the relationship between A, T and G is needed. The height, A, and the slope, T, of the forward oil price curve are in constant motion. G is an important component of the oil price, but its value is generally less volatile than A and T. The grade differential, G, is the difference between the price of the benchmark grade used to establish the value of A and T and the price of alternative grades like Bonny Light or Es Sider or Arab Light or Gippsland. The value of G moves in response to a wide range of factors that will be examined in Chapter Five. For this Chapter we will leave the value of G to do its own thing.

There is no definite relationship between the height and the slope of the forward oil curve, although there is over time a 90% correlation between the two[35]. Generally contango, i.e. the forward curve slopes up from left to right, tends to be associated with low prices. Backwardation, i.e. the forward curve slopes down from left to right, tends to be associated with high prices. But there are occasions when backwardation occurs when the absolute price is very low and other occasions when contango occurs when the absolute price is very high. Extreme backwardation or contango is often observed to precede a sharp move in the absolute level of the curve.

The natural shape of the forward curve for any commodity including oil is, *in theory*, contango, i.e. sloping up from left to right. This is because oil for delivery today can be turned into oil for delivery in, say, three months' time by placing it in storage. The only difference between the price of oil for prompt, i.e. immediate, delivery and the price of oil for delivery in three months' time ought to be the cost of leasing storage tanks and the time value of money[36]. Oil delivered today must be paid for in 30 days' time. If it is brought out of storage in three months' time and sold back into the market, payment for it will be received 30 days after that. So cash is tied up in the oil for a three month period and this must be financed.

*In practice* backwardation is not an unusual occurrence in the oil market. This is partly because oil is a very geo-political commodity subject to headlines that send oil companies scrambling to secure additional supply at short notice. This is compounded by the fact that many companies operate a "just-in-time" stock management policy that gives them a limited cushion to absorb unforeseen short notice demand.

All three price components, A, T and G, can move independently of each other, but they behave as if they are attached by a piece of elastic, which allows them some freedom of movement, but pulls them back into line if they get too far outside a logical inter-relationship with each other. This realignment process occurs through a process of arbitrage.

# Arbitrage

Arbitrage can be defined as a form of trading that attempts to profit from discrepancies in prices due, for example, to location, illiquidity, slow communication of new information, or any other reason. Typically, price imperfections are small and transient, because

---

[35] Given the large size of the sample of data we have tested we consider this relationship to be significant and moderately strong, but not determinant.

[36] There may be other second order costs such as loss of oil in-transit through tank clingage, contamination risk, loss of light ends, measurement and inspection costs etc.

traders quickly recognise and exploit any such inconsistencies. Traders will sell those prices, or price components, that are over-valued, driving the price down, and buy those price components that are undervalued, driving the price up, so that they rapidly come back into line.

The three components, A, T and G, can either be agreed individually by traders negotiating the price of oil, or they can be agreed as a composite number, i.e. a fixed number representing A+/-T+/-G.

In the case of long-term sales contracts for a number of cargoes, it is frequently the case that one, two or all three of the components of the price are set by reference to actual published prices averaged over the whole month of lifting.

In the case of spot sales, i.e. one-off sales of an individual cargo, the three components may be set by actual prices as published on the 3-5 days around or after[37] the B/L date. Or they may be fixed on the transaction date, based on market expectations at that time of what the three price components will turn out to be on those 3-5 days.

The price of physical oil can be expressed in a variety of ways by traders as they conclude their deal. For example:

- It may be fixed and flat, i.e. "$X/bbl". This is rather uncommon[38];

- It may be floating, e.g. an average of the published price of the grade (assuming that grade is assessed by one of the publications) on the 3-5 days around or after the B/L date;

- It may be partly fixed and partly floating, e.g. the average Dubai price for the month of delivery as quoted in a specific publication +/- a fixed differential agreed upfront;

- It may be as complicated or as simple a formula as the traders choose to construct.

Typically the three components of price emerge when traders are negotiating the price of a cargo of oil for delivery in the spot market.

For example, say that on 20th October a producer has a cargo of Bonny Light oil for delivery in the date range, or laydays, 17th - 19th November for sale. The producer will

---

[37] Custom and practice shows regional variations with some favouring a 3 day average and others a 5 day average. Similarly some favour an average price of which the B/L date forms the mid-point and others favour an average price based on average commencing on the day after the B/L date

[38] The exception is the pricing of benchmark grades in forward, futures, swaps and options contracts.

first locate a refiner or trader in the market that has an interest in buying the cargo.

When negotiating the price of the cargo the traders may look at the value of the absolute price, A, in the market by considering the price of, perhaps, Brent futures or 25-Day BFOE probably for December or January delivery. Bonny light cargoes are typically priced by reference to the Brent benchmark. The reason for looking at a December delivery contract as the very earliest contract month to be used to establish the value of A is that on 20th October the first contract in both the futures and the forward markets is the December contract. This was discussed in the previous Chapter Three under the heading "Which Month?"

Hence the traders can easily establish the current value of December delivery Brent, the value of A, by looking at their ICE screen or talking to a broker in the 25-Day BFOE market. This may be, say, $120/bbl. This constitutes the value, A, in our example.

But our hypothetical cargo is not for delivery in December. It is for delivery in the 17th-19th November date range. So the traders need to consider the price difference, T, between Brent for delivery in December and Brent for delivery 17th-19th November. The market that provides the price of oil for delivery on the 17th-19th November dates relative to the price of oil for delivery at any time during the month of December is the CFD market.

## The CFD Contract

CFD stands for "Contract-for-Difference", which in most commodity markets is a generic term for a swap contract. In the oil market CFD means a very particular kind of swap. It is a swap that fixes the value of the price differential between oil for delivery in a specified date range and oil for delivery in a specified forward month. The swap is cash-settled at the price differential between oil for delivery in a specified date range and oil for delivery in a specified forward month as published by a price reporting agency, usually Platts, during the specified date range of the swap. CFD market prices are quoted in delivery weeks. We will examine this market in detail below.

Hence the traders can establish very easily the value of T by talking to a broker in the CFD market. So returning to our example of the Bonny Light cargo being traded on 20th October, they can then determine the current value of Brent for 17th-19th November delivery, by adjusting the absolute December price, A, by the time differential, T. For illustrative purposes let us say that the value of T is December minus $0.75/bbl on 20th October, i.e. it is anticipated on 20th October that the price of Dated Brent as published in the week of which the 17th-19th November dates form a part, will be $0.75/bbl less than the December delivery 25-Day BFOE contract in the week of which the 17th-19th November dates form a part.

This -$0.75/bbl valuation for T is in effect stating that on the day the Bonny Light cargo is traded (20th October), market participants are willing to buy or sell a CFD contract that anticipates that the average value of Dated Brent as published on the dates of 17th-19th November will be $0.75/bbl below the average value of December 25-Day BFOE as published on the dates of 17th-19th November. Thus, in this hypothetical example, the price for Brent for 17th-19th November delivery would be $120 minus $0.75/bbl =$119.25/bbl.

But the cargo in question is not 17th-19th November delivery *Brent*. It is 17th-19th November delivery *Bonny Light*. So the traders need to consider the price difference between Brent and Bonny Light, i.e. the value of G. This is the least visible of the three components of price and will be negotiated in the context of the Brent/Bonny Light price differential most recently negotiated by other Bonny Light sellers. Or it may be set by reference to the Brent/ Bonny Light price differential quoted by a publication. This may be, say, Dated Brent plus $1.50/bbl[39]. So the "correct" price for the cargo becomes $120/bbl (A) minus $0.75/bbl (T) plus $1.50/bbl (G) =$120.75/bbl.

If on 20th October when the two traders enter into a physical contract to buy and sell the cargo of Bonny Light for delivery on 17th-19th November they actually wanted to agree a fixed and flat price, the price they could agree at the instant in time when the value of A +/- T +/- G were assessed ($120/bbl -$0.75/bbl +$1.50/bbl) would be $120.75/bbl. On 20th October these would be the values of A, T and G relevant to oil loading 17th-19th November date range.

## What is the Right Time to Set the Price?

It may be that the two parties to the deal may not wish to set these values on 20th October because 20th October is simply an arbitrary date on which the two parties encountered each other in the market. Each will have a view, possibly even opposing views, about what the values of A, T and G relevant to oil loading 17th-19th November date range are likely to be in the days and weeks between 20th October and 17th-19th November. Each may consider that their commercial interests would be best served by fixing the value of A and/or T and/or G on a different day, not on 20th October.

In these circumstances the parties could, as they used to do in the olden days of the

---

[39] We will examine the value of G in detail in Chapter Five.

early 1980s, decide to delay the execution of the contract until they find a day that suits both their respective price views. But that day might never arrive and sooner or later the nerve of one or the other might have to crack before the seller is obliged to nominate a vessel to avoid a failure to lift at the loading terminal, or the buyer is obliged to acquire crude oil at any price to avoid running out of feedstock at the refinery.

Fortunately the evolution of price risk management tools in the oil market has meant that this inefficient waiting game is no longer necessary.

As discussed in earlier Chapters, market convention dictates that the "correct" price of a cargo is its value as published on the 3-5 days around or just after its bill of lading (B/L) date, taking into account regional variations. This means that the contract can be executed with this generic formula price clause included and both parties can be assured that they have agreed a fair market price. Either or both parties can now unbundle that price formula and manage its components independently as it sees fit. This can be done by hedging the physical contract value of A in the forward or futures markets and hedging the physical contract value of T in the CFD market.

In practice, if one or other counterparty to the deal does not have in-house trading or hedge capability or authority, then the other counterparty will often comply with the price wishes of the company with less market know-how. The dominant counterparty in this situation may not wish to fix the value of A and/or T on the day the deal is executed. But if the other party with less market access does, the proficient trading company will often fix the price under the physical sales contract even though it may not like today's fixed prices for oil on the relevant delivery date. This is because it has the trading capability to take that unwanted fixed price in the physical contract and convert it into the formula price it actually wants by hedging. It can then convert this formula price back to a fixed value later at a time of its own choosing, not 20th October. The same principle applies mutatis mutandis if the less trading-focused company wants to agree a formula price and the other counterparty would prefer a fixed price[40].

As between the two counterparties to the physical contract, the price will always be whatever was agreed at the time the deal was struck, i.e. either a fixed price or a formula price. But either or both parties may choose to hedge that physical contract price with offsetting deals that deliver the overall financial result it wants. The ability to hedge actually separates the economic pricing outcome of a sale and purchase decision from the identification of an outlet for a physical cargo. So long as the fixed price or the formula price represents a fair value of oil for delivery in the relevant delivery date

[40] There may be a small charge for the service reflecting dealing costs and brokerage fees in the market.

range at the time the deal is struck, then companies with hedge capability are largely indifferent to whether the physical cargo price is fixed or floating.

It is much more difficult to hedge the value of G because there are only a very limited number of markets in which the value of G can be managed. Those instruments that do exist to manage G tend to be limited to markets in the price differentials between major benchmarks, such as the differential between Brent and WTI or Brent and Dubai. There is also a swap market in the Dated Brent/ Urals blend differential. This is of little assistance if the grade that is being traded is, say, Tunisian Ashtart, the price of which is not even assessed by the main price reporting agencies.

If the counterparties to the deal can agree the value of G at the time the deal is struck then this fixed value will be the value of G that is plugged into the A +/- T +/- G price formula. If it cannot be agreed *and the value of G is assessed by price reporting agencies* then the value of G in the price formula will typically use the value of G as published on the 3-5 days around or just after the B/L date. If the value of G cannot be agreed *and the value of G is not assessed by price reporting agencies*, this can be a deal breaker. The value of G will be discussed in more detail in Chapter Five.

## What Really Happens when Two Traders Transact A Deal

It should perhaps be made clear that the process of agreeing the price of a physical cargo described above is in reality much compressed in the actual phone call between two traders. Traders rarely, if ever, spell out in detail the values of A, T and G implied by their discussion.

For example, the deal discussion described above for the sale and purchase of a cargo of Bonny Light for delivery 17th-19th November might actually be a very short one. The parties may simply agree that the price of the cargo will be the average published price of Bonny Light on the 3 days after the B/L date. But this short conversation is actually an agreement to float the values of A, T and G over the 3 days after the B/L date. Bonny Light is an actively traded grade whose price is assessed by several price reporting agencies.

Either or both parties may wish to manage this price. For example, the buyer may wish to ensure that it has a profit when it re-sells the cargo to a third party or when it refines the cargo into oil products. At that point the values of A, T and G crystallise as the trader unbundles as much of the price formula contained in its physical purchase contract as is necessary, and as is possible, to lock in its re-sale margin or refining margin. At this point it can manage the value of A implied by the formula price in the physical contract–the price of a forward or futures contract on the 3 days after the B/L date. It can also manage the value of T implied by the formula in the physical contract – the differential between the price of wet, or Dated, oil for delivery on the 3 days after the B/L date and the price of that forward or futures contract that it has used to manage the value of A on the 3 days after the B/L date.

The time has come to look at the value of T as expressed by the CFD market in more detail.

## The CFD Market and the Value of T

The value of T describes the price at which an individual identifiable cargo of a grade of crude oil can be bought or sold today for delivery during a specific loading date range. As discussed in Chapter One, typically physical cargoes are scheduled on a monthly cycle and are traded about 10-45 days in advance of their loading date range. In the case of the Brent market they are traded 10-25 days in advance of their loading date range. At any given point in time a cargo of a particular grade of oil will trade at one price if it is scheduled to be delivered on one day and an entirely different price if it is scheduled to be delivered on different day.

The CFD market establishes the price of cargoes with specific date ranges from today's date up to 8 weeks in the future. Beyond this 8 week forward time period the slope of the forward price curve is evident in the price differential between the prices of oil for delivery in different 25-Day BFOE forward contract months or the prices of oil for delivery in different futures contract months. We will return to the back end of the forward oil curve later in this book, but for the moment we will focus on the slope of the curve up to 8 weeks in the future, i.e. in the time period covered by the CFD market.

The size of T, the slope of the forward oil curve, is determined by microeconomic supply and demand factors and stock holding levels. The value of the time differential, T, increases the more the delivery date of an individual cargo diverges from the delivery month used to assess the absolute price level, A, in the sales contract for that cargo.

CFDs are financial instruments used to hedge the very front end of the forward price curve. There is no physical delivery involved, or indeed possible, because the market is based on price differentials that are by definition undeliverable. The values expressed in the CFD market are tied directly to the physical delivery of oil *in the North Sea*. It represents the price differential between Dated Brent and the price a forward 25-Day

BFOE contract, typically the two months forward 25-Day BFOE contract[41]. A similar contract in the price differential between Dated Brent and the price of a Brent futures contract also exists. This is the dated-to-front-line (DFL) market. The two markets work exactly the same way, except that the DFL market does not trade in weekly blocks like the CFD market. It is cash settled by reference to monthly, quarterly or sometimes annual price averages.

The CFD market is very actively traded by producers, refiners and traders through brokers. Two main publications, Argus and Platts, publish end of day settlement values for this market. The CFD market reflects the difference in the price of cargoes of oil for delivery in different weekly date ranges. Market participants use the Platts published prices for the two components of the CFD, i.e. the Dated Brent price and the 25-Day BFOE forward price actually quoted during the CFD week in question, for cash settling the CFD transactions. Let's consider a hypothetical example of the market on 15th July Year X. The data that can be collected from publications such as Argus and Platts and from brokers in the market may show something along the lines of the data outlined in Table 4.

---

[41] Platts proposed at a workshop in October 2012 that this be changed to the three months forward 25-Day BFOE contract.

**Table 4 CFD Market Data**

| TODAY'S DATE IS MONDAY 15th JULY | | | | | | | | |
|---|---|---|---|---|---|---|---|---|
| CFD Weeks Trading | Week 1 | Week 2 | Week 3 | Week 4 | Week 5 | Week 6 | Week 7 | Week 8 |
| CFD Week Cargo Date Range | 15-19 July | 22-26 July | 29 July to 2 August | 5-9 August | 12-16 August | 19 -23 August | 26-30 August | 2-6 Sept-ember |
| CFD Price $/bbl | Sep. 25-Day BFOE+ 1.5 | Sep. 25-Day BFOE+ 1.25 | Sep. 25-Day BFOE + 1.1 | Oct. 25-Day BFOE+ 1.5 | Oct. 25-Day BFOE+ 1.45 | Oct. 25-Day BFOE+ 1.4 | Oct. 25-Day BFOE+ 1.35 | Nov. 25-Day BFOE+ 1.8 |
| Aug. 25-Day BFOE $/bbl | 101 | | | | | | | |
| Sep. 25-Day BFOE $/bbl | 100 | | | | | | | |
| Oct. 25-Day BFOE $/bbl | 99.5 | | | | | | | |
| Nov. 25-Day BFOE $/bbl | 99 | | | | | | | |
| By simple Arithmetic: CFD Price $/bbl | 101.5 | 101.25 | 101.1 | 101 | 100.95 | 100.9 | 100.85 | 100.8 |

All of this data is collected at one instant in time on 15th July. At that time the CFD market is trading 8 weeks forward. On 15th July, week 1 represents the trading week of 15th-19th July, i.e. very prompt cargoes. Week 2 represents the trading week of 22nd-26th July; week 3 the 29th July to 2nd of August and so on out to week 8 which represents the week of 2nd-6th September.

To recap on physical logistics, on 15th July oil producers will know the dates of their cargoes loading all throughout August but few, if any, will yet have been informed by their respective terminal operators which loading date ranges will be allocated to their oil cargoes for loading in September.

On 15th July the CFD market for week 1, the 15th-19th July, is trading at September 25-Day BFOE plus $1.50/bbl. The CFD market for week 4, the 5th-9th August is trading at October 25-Day BFOE plus $1.50 /bbl. The CFD market for week 8, the 2nd-6th September, is trading at November plus $1.80/bbl. The fact that September cargo loading dates have not yet been issued by the terminal operators does not prevent traders from having a view on the value of dates for cargoes being delivered in early September compared with those being delivered in late September.

However on 15th July there is a liquid market, not in individual cargo dates for September but for 25-Day BFOE for delivery in September, October and November. So the fact that the CFD market for week 1, the 15th-19th July, is trading at September 25-Day BFOE plus $1.50/bbl allows us to calculate that the value of the 15th-19th July dated week is $101.50/bbl, i.e. $100 +$1.5/bbl, on 15th July.

Similarly the fact that the CFD market for week 4, the 5th-9th August, is trading at October 25-Day BFOE plus $1.50/bbl allows us to calculate that the value of the 5th-9th August dated week is $101/bbl, i.e. $99.50 +$1.5/bbl, on 15th July.

Also the fact that the CFD market for week 8, the 2nd-6th September, is trading at November plus $1.80/bbl allows us to calculate that the value of the 2nd-6th September dated week is $100.80/bbl, i.e. $99 +$1.80/bbl, on 15th July.

A word of warning about the CFD market: the publications that report the prices of CFD weeks are not consistent on the forward month they use to report the value of the CFD week differentials. However so long as it is clear the size of the differential that applies and to which forward month that differential should be added or subtracted then this should make no difference. For example, if the 15th-19th July week is trading at September 25-Day BFOE plus $1.50/bbl and the September/October inter-month price differential is $0.50/bbl, then the 15th-19th July week could be equally correctly be reported as trading at October plus $2/bbl.

A further word of warning: the hypothetical data we have produced for this illustrative example fits together quite neatly and shows a forward oil curve in consistent backwardation (See Figure 23) out to the November 25-Day BFOE contract. In reality the forward curve, particularly in the first four weeks is rarely so linear and so consistent with the first month forward 25-Day BFOE contract. Anomalies occur that can provide arbitrage opportunities for traders.

For example, take the August 25-Day BFOE contract in our example above. Any producer of Brent, Forties, Oseberg or Ekofisk for delivery in the last few days of August would be well advised to sell this cargo as a 25-Day BFOE August cargo, rather than sell it as a Dated Brent or Dated Forties or Dated Oseberg or Dated Ekofisk cargo with a late August loading date. This is because the last week of August is trading at $100.85/bbl in the CFD market, but at $101/bbl in the August 25-Day BFOE market.

**Figure 23 The Forward Oil Curve 15 July Year X, Showing CFD weeks and 25-Day BFOE Market**

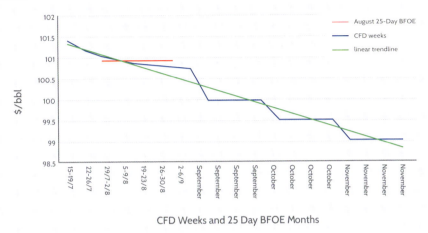

However lifting schedules at each of the B, F, O and E terminals are extremely transparent. So if any producer entered the market to sell a 25-Day BFOE cargo after the lifting schedule for August was published, buyers would be aware that such producer owned a late August physical cargo and was likely to deliver it in satisfaction of an August delivery 25-Day BFOE commitment. Buyers would adjust their 25-Day BFOE purchase price ideas accordingly for that particular producer, particularly if it was a company that entered the market only infrequently.

## Fixing the Value

To demonstrate how this data is deployed in the market, consider an example where two traders were discussing the price of a cargo for delivery in the week of 5th-9th August on 15th July. Table 4 tells us that on 15th July the value of a cargo of Brent for delivery within the date range of in the 5th-9th August delivery week was about $1.50/bbl more than the value on 15th July of a cargo of 25-day BFOE for delivery in October. So if the counterparties to a physical deal in oil for delivery in the week of 5th-9th August had wanted to fix the value of the absolute price, A, and the time differential, T, on 15th July 2011, the fixed price they would have agreed would have been about $99.50/bbl plus $1.5/bbl = $101/bbl. This calculation is based on the mid-point of the bid-offer negotiating range over which the two counterparties would haggle before executing the deal.

If the grade of oil being transacted was not Brent the counterparties would agree the fixed value of any relevant grade differential, or would agree a formula for calculating that value of G, at the time the deal was struck.

# Floating, Fixing or Hedging in Three Parts

**Figure 24 Crude Oil Price Components**

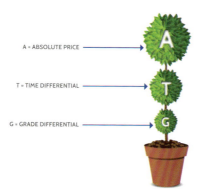

A = ABSOLUTE PRICE

T = TIME DIFFERENTIAL

G = GRADE DIFFERENTIAL

Instead of fixing the value of the cargo for delivery 5th-9th August on 15th July the counterparties may have agreed to float the price over the 3-5 days around or just after the actual B/L date of the cargo, whatever that turned out to be when the price assessments were published at the end of each day. If they decided to float the value in their physical contract, the data contained in the above table ceases to have any relevance to them. If the parties do not lock in a value when they see it that value cannot be retrieved later when the market has changed and both the absolute level, A, of the forward oil curve and the slope of the forward oil curve, T, have shifted to different levels.

If on 15th July the counterparties to the physical deal for the cargo for delivery in the 5th-9th August had decided to float the cargo's price over the 3-5 days around or just after the actual B/L date then the outcome of the price formula in the physical contract would be determined by calculating the average differential between the values of Dated Brent and 25-Day BFOE published on those 3-5 days around or after the B/L date. If they decided to use this floating price formula each could separately unbundle the price and manage it by hedging. Any such hedging action would only have relevance to the company undertaking the hedge and would have no impact on the other counterparty to *the physical transaction*.

Consider further the case where the two counterparties to the deal agreed to float the cargo's price as the average of the 3-5 days around or just after the actual B/L date. If we assume that the seller of the physical cargo for delivery in the 5th-9th August date range had a view that the market value of T was about to move against it[42], e.g. that the value of the time differential, T, was about to fall from October 25-Day BFOE plus $1.50/bbl to perhaps October 25-Day BFOE minus $0.25/bbl, then it would want to offset the floating formula price in the contract by hedging. To do this it would sell the 5th-9th August CFD week at the fixed price available in the market on 15th July for that 5-9 August CFD week, i.e. by locking in its value at October 25-Day BFOE plus $1.50/bbl.

After the close of business on 9th August the two counterparties to the physical trade that had agreed to float the value of T over the 5-9 August week would be able to work out the physical contract price.

Let's say the average published differential between the price of Dated Brent and the price of October 25-Day BFOE over the five days of 5th, 6th, 7th, 8th and 9th August had turned out to be minus $0.25/bbl, as the seller had anticipated. The physical contract price would reflect the value of the time differential, T, as minus $0.25/bbl. Hence the seller would have achieved a lower price than it would have achieved if it had persuaded the buying counterparty to the physical deal to fix the value of T in the physical contract at plus $1.50/bbl on 15th July when the deal was struck.

However this hypothetical opportunity cost on the physical contract would be offset by a gain on the CFD hedge. The seller sold the CFD at a fixed price of October 25-Day BFOE plus $1.50/bbl on 15th July. For the seller of the physical cargo, this CFD hedge would cash settle by buying back at the average published differential between the price of Dated Brent and October 25-Day BFOE over the five days of 5th, 6th, 7th, 8th and 9th August, which, as we have said above might have turned out to be minus $0.25/bbl. So the physical cargo seller would have sold the CFD at plus $1.50/bbl and bought it back (cash-settled) at minus $0.25/bbl. This would deliver a hedge profit of $1.75/bbl.

For the seller of the physical cargo the overall result would be that it sold the physical cargo with a price component, T, of minus $0.25/bbl to which it could add a hedge profit of $1.75/bbl. This would deliver a net result for the price component, T, of minus $0.25/bbl plus $1.75/bbl = plus $1.50/bbl. The seller of the physical cargo has achieved exactly the same result as it would have achieved if it had persuaded its physical cargo buyer to fix the value of T in the physical contract price formula on 15th July, as it wanted

---

[42] It might also have had a view on which way the value of A was likely to move and may have wished to hedge that too. If it did it could do so by entering into a separate hedge of the value of A in the futures or forwards contracts, as described in Chapter Three.

to do. This is a successful hedge.

Alternatively, let's say the average published differential between the price of Dated Brent and October 25-Day BFOE over the five days of 5th, 6th, 7th, 8th and 9th August had turned out to be plus $2/bbl, contrary to the seller's expectation. The physical contract price would reflect the value of the time differential, T, as $2/bbl. Hence the seller would have achieved a higher price than it would have achieved if it had persuaded the buying counterparty to the physical deal to fix the value of T in the physical contract at plus $1.50/bbl on 15th July when the deal was struck.

However this hypothetical better-than-expected outcome on the physical contract would be offset by a loss on the CFD hedge. The seller sold the CFD at a fixed price of plus $1.50/bbl on 15th July. For the seller of the physical cargo this CFD hedge would cash settle by buying back at the average published differential between the price of Dated Brent and October 25-Day BFOE over the five days of 5th, 6th, 7th, 8th and 9th August, which, as we have said above might have turned out to be plus $2/bbl. So the physical cargo seller would have sold the CFD at plus $1.50/bbl and bought it back (cash-settled) at plus $2/bbl. This would deliver a "hedge loss" of $0.50/bbl.

For the seller of the physical cargo the overall result would be that it sold the physical cargo with a price component, T, of plus $2/bbl to which it would have to deduct a hedge loss of $0.50/bbl. This would deliver a net result for the price component, T, of plus $2/bbl minus $0.5/bbl = plus $1.50/bbl. The seller of the physical cargo has achieved exactly the same result as it would have achieved if it had persuaded its physical cargo buyer to fix the value of T in the physical contract price formula on 15th July, as it wanted to do. This too is a successful hedge. The seller may wish it had not hedged because its market view turned out to be wrong. But the hedge gave the seller of the physical cargo the certainty it wanted of achieving a sales price component, T, for the cargo of plus $1.50/bbl.

As mentioned in Chapter Three above, hedging does not necessarily deliver the best price outcome: its job is to give the buyer or the seller the certainty of knowing its economic position well in advance.

# Hedging and the Slope of the Forward Oil Curve

One of the perennial complaints of CEOs and CFOs of companies that are potential oil hedgers is that when prices are high and they want to enter into a large scale forward sale, the market is in backwardation. So the price at which they are able to sell is considerably below the prompt prices that are catching the headlines in the Financial Times or the Wall Street Journal. It takes a brave CFO to sell forward for 1-2 years at $120/bbl when the price for immediate delivery is $145/bbl.

Similarly when prices are low and refiners and other buyers such as airlines want to enter into a large scale forward purchase the market is in contango. So the price at which they are able to buy is considerably higher than prompt prices. It takes an equally brave CFO to buy forward for 1-2 years at $80/bbl when the price for immediate delivery is $65/bbl.

Theoretically, there is a way out of this conundrum: spread hedging.

For example, if the price of oil fell to $75/bbl this might well be because there is a surfeit of oil in the short term pulling the whole market down. But this may be associated with contango in the forward curve whereby the price of oil for delivery one year forward is $80/bbl. A producing oil company may not be tempted to hedge at $80/bbl. However it may find the $5/bbl of contango between the price of oil for delivery in the current year and the price of oil for delivery next year forward to be attractive. It could capture this contango *by buying the year 1/ year 2 forward spread, T, at -$5/bbl*[43]. In other words it would buy the differential between the first year price and the second year price at minus $5/bbl, i.e. buy year 1 and sell year 2.

Let's assume that time passes and in due course the price for immediate delivery rises to $110/bbl, perhaps because of a political conflict in an oil producing region. This could well be associated with backwardation. Say the price for delivery one year forward is $105/bbl.

At this point the producing oil company might well be tempted to hedge the absolute price A by selling one year forward, but may find the $5/bbl of backwardation to be off-putting. If it had previously bought the year 1/year 2 forward spread at -$5/bbl it could now sell it at +$5/bbl, taking a $10/bbl profit. This would in effect boost the price at which it could sell forward the absolute price A from $105/bbl to $115/bbl. But beware. This needs constant review and management.

While waiting for the market to deliver a high absolute price, A, at which the producer wishes to hedge, the spread hedge may need to be adjusted as time passes. Such

---

[43] Backwardation is a positive number because it represents the subtraction of the lower forward price quote from the nearer and higher price quote. Contango is represented as a negative number because it represents the subtraction of the higher forward price quote from the nearer and lower price quote.

adjustment will involve selling the year 1/year 2 spread and buying the year 2/year 3 spread. If the market has moved into steeper contango this will incur a loss. It will be difficult to distinguish between this "hedge management loss" and a "speculative loss".

The same principle would apply in reverse for a company that would ordinarily be motivated to hedge the value of A by buying forward oil when the price is low, an oil consumer. In this case the company would sell the year 1/ year 2 forward spread, T, at +$5/bbl when the price was high and there was backwardation in the market. When the absolute price falls to a level where the oil consumer might want to hedge the absolute price, A, by buying one year forward it could then buy back the year 1/ year 2 forward spread, T, at -$5/bbl, making a profit of $10/bbl. This profit would in effect reduce the price at which it could buy forward the absolute price A from $80/bbl to $70/bbl.

In reality the hedging company could find it risky to buy or sell the "year 1/year 2 forward spread" because it would require time to pass for both the level and slope of the forward curve to change to allow its strategy to work out. Buying or selling anything "prompt" by definition means that the contract concerned is close to expiry and some action would be required to roll the position forward in time to allow the overall strategy to unfold.

In practice it would be extremely difficult for any company to "warehouse" favourable spreads in this way. Despite the actual motivation, to improve the opportunity of locking in an effective hedge price, the action required would be difficult to distinguish from speculative trading and would be very easily misunderstood if the time differential hedge made a loss. Chapter Six discusses more practical ways in which oil producers and consumers may wish to carry out strategic hedges.

This Chapter has examined the slope of the forward oil curve and dealt with the question of valuing and hedging the price of oil for delivery in different time periods. The question of the value of T is one with which even some experienced physical traders still struggle, particularly in regions of the world where active risk management is not the norm.

As discussed above the values of A, T and G crystallise most clearly when a company decides to unbundle the price for hedging purposes. Companies that simply buy or sell on a published benchmark price formula related to the B/L date often mix and match the A, T and G components so automatically and routinely that they often give little thought to the separate value of each. This can mislead the less trading-orientated companies into believing that they are achieving a better value for G than is actually the case. What they may actually be mistaking for the value of G is a composite number combining both T and G.

# Term Contract Price Too Good to be True? Examine T!

Companies that sell under a term contract based on a monthly average published price will often achieve an apparent grade differential, G, that appears to be much higher than the published value of G might seem to justify. This is particularly the case if the buyer is allowed to choose when in each month it can load the cargoes it is purchasing. In a backwardated market, the buyer will choose early dates and in a contango market it will choose late dates. That way it can re-sell the oil it receives at much more than the monthly average price at which it has purchased the oil. In these circumstances the buyer may be prepared to pay a slightly inflated apparent value of G, because it is factoring into its thinking the benefit of choosing the loading date range of cargoes that maximise the value of T in its re-sale contracts. If the loading terminal involved is a flexible one then the seller may be giving away a much higher value of T than it is receiving in the inflated value of G.

Where the seller has to be careful in such circumstances is if it is being taxed on a price related to the B/L date rather than at the monthly average price it actually receives from the seller. The apparently good value of G, that actually includes an element of T, may be achieved at the cost of taking on substantial risk in the value of A.

For the buyer in this scenario the profit opportunity can be much greater than the slope of the forward curve. If the buyer is also a producer in the same field, buying third party oil from its JV partners can give it the opportunity to finesse the make up of each cargo to improve its revenue stream. It may be possible for it to load its own equity oil on to low priced cargoes, where it will pay oil tax at a high rate, and load non-equity third party oil onto high priced cargoes, where it will pay corporation tax at a lower rate. The taxation of crude oil cargoes will be discussed in more detail in Chapter Six.

But before we get to there we must change pace and examine the third component of the oil price, the grade differential, G. This requires us to get our hands dirty by delving in to what a barrel of crude oil actually contains.

# CHAPTER 5

## The Grade Differential, G

So far in this book we have examined in detail the first two of the three components of the oil price: the absolute price, A, and the time differential, T. In this Chapter we will look at the third component, the grade differential, G.

We stated in earlier Chapters of this book that when crude oil contracts are being transacted the traders typically specify which benchmark grade of crude oil, A, they are using as a price referencing point. We described the most common benchmarks in Chapter Three. The traders' discussion of the contractual oil price formula also takes into account the time period relevant to the benchmark price quotations being used, T. We discussed the value of T in Chapter Four.

If the grade of oil being traded is not the benchmark grade, or is loading in a different time period from the one being used to set the value of the benchmark as discussed at the end of this Chapter, a further price component will have to be included in the price formula. These adjustment factors tend to be grouped together under the heading of the grade differential, G.

The value of G is determined broadly by:

• the quality of the grade;

- supply and demand for oil products;

- the strength of the freight market; and,

- any other relevant factors.

## Crude Oil Quality

Arguably the most significant of the components of G is the quality of the grade in question compared to that of the benchmark grade.

With apologies to Forrest Gump, crude oil is like a box of chocolates. The brand name on the box – Brent, Cano Limon, Arab Light- gives us some idea of what to expect of the contents of the box, but it is not until the individual chocolates are analysed that we know "what you're gonna get". The crude oil box of chocolates is opened by the refining process. Refiners and end consumers are not interested in the brand name of the crude oil. They are interested in the quantity and quality of oil products, or chocolates, that the crude oil will yield in the refinery.

In our discussion of benchmark grades in Chapter Three we identified the various grades that are used as price reference grades solely by their API gravity and by their sulphur content. For the refiner this does not even get close to providing sufficient information to convince it to risk processing a grade of crude oil it has never tried before. Before the refiner will risk its expensive equipment by running an unfamiliar grade of crude oil, it will need to see a complete refining assay and a sample of the crude.

## The New Crude Discount

Often when a new oil field is found the explorers will take a sample of the crude and conduct a Pressure, Volume and Temperature (PVT) analysis. This will tell the geologists how the crude is likely to flow in the reservoir. Some months later, when the field development plan is being modelled, upstream colleagues will present this PVT assay, i.e. laboratory analysis of quality, to their traders and ask for an evaluation of the crude oil concerned. This data is useless to the trader. In order to understand the price at which the crude may be sold in the market, the trader will need to see a refining assay. This is because a refiner needs to know how crude oil will perform in its refinery, not how it will flow in the reservoir. This requires a full refining assay.

Oil producers testing a new field or reservoir would be well advised to take copious samples of the crude oil during the well testing stage and to have these samples analysed in a laboratory to establish the refining characteristics of the oil. If the collection of samples for refining analysis has to wait until production commences then marketing the oil will be very difficult. As mentioned above, refiners are very unwilling to risk their plant by running a full cargo of an unknown crude and will demand at the very least a full refining assay and in many cases their own samples, sometimes up to five twenty-litre samples, before buying a cargo. Samples will also have to be taken from the first production cargoes and analysed too so that potential buyers can be updated and informed of the production flow quality. But this can take weeks, sometimes months and in the meantime the market needs some data to go on for valuing the first few cargoes.

Marketing a new grade of crude oil is no different from marketing any other new product- be it shampoo, cars, or the latest i-pod. The customer wants to know what it is getting and will shop around before it commits its money. In the absence of full details the refiner may be induced to buy an early cargo of the new oil, but it will require a heavy discount to take the risk. This is referred to as a "new crude discount" and the less the information available, the larger that discount will have to be.

This is because the refiner will usually take the precaution of keeping the cargo in segregated tankage and testing it thoroughly before running it through the plant. It will often blend the new grade into its "slate" gradually, in case there are any unforeseen problems with the untested commodity. The slate is the mixture of grades of crude oil that the refinery manager and traders agree will optimise the economics of the refinery at any given time. This slate is established by an iterative process usually employing a linear programming (LP) model[44], which is regularly adjusted with new crude oil and other feedstock acquisition cost information and refined product sales price information.

Most quality problems can be blended away, i.e. diluted by commingling with known grades, but this may take time and will incur financing and storage costs for the refiner.

Producers that do not have in-house trading capability are frequently caught on the horns of a dilemma when a new field starts up. The producer wants above all else security of offtake, to avoid the failure to lift situation described in Chapter One. But the producer also wants to sell its oil at the highest price it can get.

When a shippable quantity has accrued and the storage at the loading port is approaching tank tops the producer wants to be sure that a tanker will come in to lift the oil in a safe and timely manner that does not interrupt the flow of oil. When new fields

---

[44] Not all quality attributes blend in a linear fashion.

start-up it is notoriously difficult to estimate exactly how much oil will be produced and by when a liftable quantity will have accrued. So scheduling tankers to lift early cargoes requires considerable flexibility from, and potentially involves a substantial cost to, the buyer. The buyer requires a price discount to compensate for this effort and cost. If the oilfield in question is small or has a limited life span or is located in an area off the normal tanker traffic lanes, the discount required to attract a buyer in the early stages of production will be much greater. This is because the buyer will only take the trouble to accommodate the producer's uncertain quality and uncertain timing if there is a prospect of it identifying a new grade of oil that will become a regular feature in its slate over a number of years.

The buyers best able to cope with unknown quality and doubtful timing of cargoes are major integrated oil companies that have their own refining system and storage capacity to handle uncertainty. Some of the larger trading companies have a similar level of flexibility by virtue of their contractual access to facilities that can handle short-notice cargoes of variable quality.

It is often at this stage, in choosing the buyer that will lift the first cargo of oil from a new field, that a producer takes a decision that can affect the price at which it will be able to sell its oil for years to come, or possibly over the life of the field. In its concern to secure a safe buyer that will always have a ship available to lift the oil when needed, the producer will sometimes enter into a long-term contract with a single buyer. In many cases the producer runs a tender to sell the oil to the highest bidder amongst a pool of perhaps about six or so major oil companies and trading companies.

Depending on how much quality information is available when the tender is run, third party buyers may be bidding in an information vacuum. This is unlikely to give the best price result for the seller over the long term when the production quality of the oil is established. If one of the bidders is also a producing partner in the oil field, possibly even the field operator, it will be best placed to make an informed bid on the oil quality and timing and it is likely to win the tender, without over-paying for the oil.

If the oil turns out to have attractive quality attributes the buyer has no incentive to advertise this fact because it will want to come back for additional cargoes and will not want to encourage competition. If the buyer is also a field joint venture producing partner it will have an additional incentive to under-state the value of the oil because the lower the market price of the oil, the less tax it will have to pay and the lower the price at which it will recover its costs and calculate its share of profit oil under any Production Sharing Contract (PSC).

# Important Message

There are several messages to take away from this section:

- When a new crude oil stream is being marketed the joint venture partners need as much advance information about the quality of the oil in the reservoir as possible. Samples should be taken and analysed by an independent laboratory at every opportunity, including well tests and from early production cargoes;

- This information is only valuable if it reaches the hands of potential buyers, particularly refinery managers and the traders responsible for refinery feedstock acquisition. Refining assays and brochures should be freely available and a crude oil marketing campaign, akin to that of the launch of any other new product, may well pay dividends;

- If the operator of the joint venture is a potential buyer of the crude oil stream it has a conflict of interest and it should be prepared to demonstrate that any marketing campaign undertaken as an operator obligation under the terms of the JOA is done at arm's-length on behalf of the joint venture (JV) partners;

- While secure offtake of early cargoes is of paramount importance, any contract that ties up the production stream for a year or more before the market has had the opportunity to sample and test the new grade of oil is likely to leave money on the table in the long term.

Non-integrated oil producers, i.e. those without their own refining system, have a strong incentive to maximise the price received for early production cargoes because when production starts up the project is, by definition, at the lowest point in its cash flow cycle. However leaving a few cents on the table for the first 4-10 cargoes, depending on size, may pay dividends for the field in the long term by allowing as wide a range of potential buyers as possible to run test cargoes.

# The Refining Assay

Up until the early 1990s integrated oil producers who planned to run their own oil production, and that of their joint venture partners, in their own refineries sought to keep the quality of oil field production confidential, protected by the confidentiality provisions of the joint venture operating agreement (the JOA). This is no longer the case and any JV operator that is today resistant to launching a full marketing campaign on behalf of its JV partners should be treated with some suspicion.

Most operators publish crude oil assays on their own, or on the JV's, website and often produce a brochure describing the logistics of loading oil at the relevant terminal, i.e. the terminal regulations, jetty sizes, ballast handling, storage capacity, etc. The assay describes the handling characteristics of the whole unrefined crude oil and also describes the quantity and quality of oil products that can be derived from the crude oil in the refining process.

To analyse fully a refining assay requires the services of a chemical engineer, but we include in the ensuing pages a few pointers for use by the layman.

In very general terms and with plenty of exceptions, the "lighter" the crude and the lower its sulphur content, the higher the price it will attract in the market. Crude oil is categorised as "light" or "heavy" according to its API[45] gravity, which is a measure of the density, or specific gravity, of a substance relative to that of water. The formula for calculating API gravity is:

$$°\text{API Gravity} = \left( \frac{141.5}{\text{Specific Gravity}} \right) - 131.5$$

The API gravity of water is 10°. Various grades of crude oil have gravities ranging from 15° API to 45° API or more, although some South American grades have an API of <10°. The higher the API number, the lighter the crude oil. Once the API gravity exceeds 45°, the material is generally referred to as condensate, rather than crude oil.

Sulphur is a contaminant which is present to a greater or lesser extent in all crude oil. There are no hard and fast rules, but crude oils with less than 0.5% sulphur by weight are typically referred to as low sulphur or "sweet". High sulphur crude oils have sulphur contents > 2% and are referred to as "sour". Between 0.5-2% sulphur content can be referred to as medium sulphur.

The term "sour" can also refer to the presence of sulphur compounds such as hydrogen sulphide ($H_2S$) or mercaptans. $H_2S$ is a highly poisonous gas on which very strict limits are placed by pipeline operators receiving the crude oil. Mercaptans are the compounds that are added to natural gas to give it a rotten egg odour, which makes it easier to detect gas leaks. High mercaptan crudes can present shipping difficulties as the odour can remain in the tanks for several cargoes and can cause complaints from towns close to the refinery processing it.

---

[45] American Petroleum Institute

Environmental legislation to burn cleaner fuels means that sulphur compounds need to be removed in the refining process. The cost of removing sulphur depends where in the refining process the sulphur compounds are concentrated and in what precise form of compound they occur.

The Total Acid Number (TAN) is the amount of potassium hydroxide in milligrams required to neutralise a gram of crude oil and it is therefore a measure of the acidity of the whole crude. A TAN of >0.3 can present difficulties for refineries that do not have acid resistant metallurgy and indicates that the refinery may need to blend the crude oil with a low acid grade. High salt content can also be corrosive and require treatment. The presence of metals such as vanadium and nickel can cause problems for refineries that use catalysts because the trace metals can deposit on the catalyst surface slowing down the reaction time.

Other key whole crude properties that can give an indication of the handling characteristics of the oil are its pour point °C, its wax content % weight and its kinematic viscosity. These factors will indicate how the crude will flow at ambient temperature and indicate whether transporting the oil will require the use of heated tankers or a pour point depressant. Such factors will also determine whether the crude oil is suitable for transport by pipeline.

To examine crude oil properties in any further depth requires a bit of basic organic chemistry, which we will cover briefly in the next sub-section. Readers with no interest in these matters may safely skip this sub-section.

## Paraffinic and Naphthenic Crude Oil

Traders encountering a new crude for the first time will often enquire if the oil is paraffinic or naphthenic. This sub-section will explain what is meant by these terms.

An atom is made up of a positively charged nucleus surrounded by negatively charged electrons. The number of positive charges in the nucleus determines how many electrons are "needed" to stabilize, or neutralize, the atom. The valence of an atom refers to the number of negative electrons it has in orbit in the outer electron shell round the positive nucleus. The atom will tend to gain or lose electrons in order to neutralize the charge of its nucleus.

The valence of an element is related to its ability to combine with hydrogen (H), which has a valence of 1, to achieve neutrality. For example, one oxygen atom combines with two hydrogen atoms to form water and, since the valence of each Hydrogen is 1, the valence of oxygen is thus determined to be 2. An atom of Carbon (C) is capable of combining with up to four other atoms, i.e. it has a valence number of 4.

So one familiar basic hydrocarbon molecule is $CH_4$, i.e. methane. This represents one C, with a valence of 4, combined with 4 H, each with a valence of 1.

$$H \underset{\underset{H}{\overset{\overset{H}{|}}{|}}{-}}{\overset{\overset{H}{\underset{|}{}}}{}} C - H \quad \text{i.e. Methane} = CH_4$$

Carbon atoms can combine not only with atoms of other elements, like hydrogen, but with other carbon atoms. This means that carbon atoms can form chains and rings onto which other atoms can be attached. Carbon compounds are classified according to how the carbon atoms are arranged and what other groups of atoms are attached. "PONA" indicates how carbon chains or rings are organized. PONA stands for Paraffins, Olefins, Naphthenes and Aromatics.

**Paraffins,** also known as alkanes, are straight (normal) or branch (iso-) chained hydrocarbons "saturated" with hydrogen. In other words the 4 - valence number of the carbon has been neutralized by the attachment of sufficient 1- valence hydrogen atoms to use up all the valence of the carbons.

Paraffinic material is used for making ethylene and propylene, which are the building

$$H-C-C-C-C-C-H \quad \text{i.e. normal or n-pentane} = C_5H_{12}$$

blocks to make polythene and polypropylene in the petrochemical industry. Paraffins occur in all crude oils, especially in so-called paraffinic crude in the lightest distilled fractions.

By referring to alkanes or paraffins as "saturated" we mean that there are no double or triple bonds between the carbon atoms, i.e. the valency is "neutralised" because each 4 valency carbon atom is attached to four other atoms, some of which may be other carbon atoms.

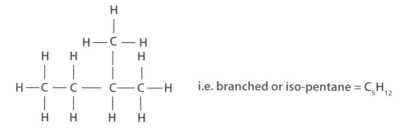

i.e. branched or iso-pentane = $C_5H_{12}$

**Olefins,** also known as alkenes, are unsaturated and are made up of hydrocarbons containing carbon double bonds. When there are insufficient hydrogen or other atoms available, the 4-valency of the carbon atom is "unsatisfied". The carbon will attempt to "acquire" a spare electron from another carbon atom. The carbon atoms are depicted as sharing the available short supply of electrons amongst themselves and forming double bonds.

Olefins tend not to occur naturally in hydrocarbons because the double bonds are highly reactive. The olefins are quickly converted to more complex molecules where all of the carbon's "appetite" for electrons is satisfied. However, when large carbon molecules are broken up in the refining process, such as in a cracker, olefins tend to be more prevalent, i.e. carbon to carbon bonds are broken and valency needs to be satisfied by saturation with hydrogen. If there is insufficient hydrogen available double bonds (alkenes) or even less stable triple carbon bonds (alkynes) will form.

For this same reason olefins tend to poison catalysts in catalytic crackers because they are so reactive they can build up a residue of molecules on the catalyst surface. The presence of olefins in what purports to be crude oil may indicate that the material has been spiked with some refined product, perhaps with cheap cracked fuel oil.

```
              H
              |
        H ― C ― H
  H   H       |   H
  |   |       |   |
H ―C ― C ―― C == C
  |   |       |
  H   H       H
```

i.e. branched or iso-pentene = $C_5H_{10}$

**Naphthenes,** also known as cyclo-paraffins or C-alkanes, are saturated cyclical chains of more than 4 carbon atoms, such as cyclopentane or cyclohexane, the latter depicted below.

Cyclohexane

**Aromatics** are cyclical unsaturated hydrocarbon chains such as benzene, toluene and xylene. Naphthenic and Aromatic (N+A) ring structures are typically used in petrol production and aromatics are used in making polystyrene, paint, solvents etc.

Benzene

The simplest aromatic, benzene, is depicted above. It is often represented as a carbon ring with a cloud of electrons in the middle shared amongst the six carbons.

## Benzene

The proportion of the four types of hydrocarbons - PONA - in crude oil sum to 100% and can affect its usefulness for specific purposes. For example, crude producing high paraffinic, light naphthas are used for petrochemical plants, whereas heavier naphthas with high naphthenes and aromatics are typically used in gasoline production. Aromatics are used in making polystyrene and paint solvents. The proportion of these types of hydrocarbons in other oil products such as gasoil has less relevance. However other product specifications not generally linked to the PONA proportions are more significant.

## Refining Processes

The refining assay provides data on the PONA content of crude oil. It also gives an indication of how the crude oil will perform in a refinery. Refining processes can be classified into three basic types:

- Separation;

- Treatment; and,

- Upgrading/Conversion.

**Separation:** Crude oil is a mixture of many thousands of different chemical compounds, all of which have their own unique properties. The separation process is employed to extract certain ranges or types of hydrocarbon compound from the mixture. The first refining process separating compounds in accordance with their different boiling points is primary distillation, which can be simulated in a laboratory using stabilised[46] samples of the oil.

The refining assay provides a True Boiling Point (TBP) distillation curve that indicates the yield of each group of products at varying temperatures up to about 300-350°C. Beyond that temperature oil begins to thermally decompose, i.e. long chain molecules start to break down. High vacuum distillation up to about 500°C can be employed when deep distillate yield information is required, but the typical assay stops short of thermal cracking. Primary distillation involves separation alone and afterwards the various products could, theoretically, be re-combined into crude oil. Separation by freezing points, precipitation, and insolvency in suitable solvents can also be employed.

**Treatment:** Treatment processes are primarily used to treat sulphur compounds contained in the refined products, or distillates. This may involve hydrodesulphurisation in which sulphur is removed by first converting the sulphur compounds to $H_2S$, which is a gas that is easily separated from liquid products. Elemental sulphur can be recovered from the gas. An alternative process is Merox treatment, which involves converting mercaptans by oxidising them into disulphides, which are sweeter and non-corrosive. Other treatment processes remove acid or metals. Treatment processes do not affect the yields of individual products, but are primarily used to meet required finished product specifications. They are chemical processes that cannot be reversed.

**Upgrading/Conversion:** The third type of process encountered in oil refineries is upgrading or conversion. These processes are employed to change the yield of products, and generally consist of converting the unwanted high boiling point, long chain hydrocarbons to shorter molecules. This is achieved through application of heat and catalysis to "crack" the molecules. Cracking is an endothermic reaction (i.e. it

---

[46] Oil from which the gaseous light ends have been extracted

absorbs heat), but also produces carbon, which can be used in the process to generate heat.

# The Gross Product Worth (GPW)

When a new crude oil is encountered in the market the refining assay will begin to provide clues as to the likely value of the new grade relative to existing benchmark grades. This is because the assay spells out the quantity and quality of finished and semi-finished products that can be derived from the grade under laboratory conditions. This gives us the starting point for evaluating the Gross Product Worth (GPW) of the crude oil.

Crude value assessments are based on simple product yields multiplied by the prices of the respective products that can be derived from it:

$$GPW = \sum ((\text{Yield of Product 1} \times \text{Price of Product 1}) + (\text{Yield of Product 2} \times \text{Price of Product 2}) + (\text{Yield of Product 3} \times \text{Price of Product 3}) \text{etc.})$$

There are a range of publications available that provide daily product price assessments for a wide range of products of varying qualities and specifications in different areas of the world. The two most commonly used in the oil industry are Argus and Platts. This data can be used to calculate the GPW of a new grade of crude oil in different regional markets, such as N.W. Europe, the US Gulf Coast, Singapore, and Japan etc. Each region has different product quality specifications depending on such factors as:

- environmental legislation, which may impose lower sulphur limits in one area compared to another;

- the use to which the products will be put. For example kerosene may be used for heating or lighting in one region as opposed to for the production of jet fuel in another region;

- the temperature range encountered in the country in question. For example, winter specification diesel will demand a much lower pour point than summer grade diesel;

- seasonal variations, which demand more transportation fuel in the summer and more heating oil in the winter, etc.

Once the most appropriate product prices for the product yield have been chosen a simple GPW can be derived. But this of itself is an incomplete picture of what the crude oil is worth because crude oil rarely sells at its GPW value. For example, the GPW of Dubai crude oil bears little relation to the market price of Dubai crude oil.

This is because every single refinery will have its own GPW model depending on what treatment and upgrading plant it possesses. So there is no single correct GPW for any given grade of crude oil.

Nevertheless the GPW calculation can begin to explore the likely differential at which a new grade of crude oil will trade in the market relative to a benchmark grade. The steps involved in estimating a GPW differential are:

- Work out the GPW of the new grade in question, i.e. $GPW_{newoil}$;

- Work out the GPW of the benchmark grade, i.e. $GPW_{benchmark}$;

- Calculate the GPW differential, i.e. $GPW_{benchmark}$ minus $GPW_{newoil}$. Let's say this is $2/bbl;

- Apply the GPW differential to the market price of the benchmark grade[47], i.e. the Market Price$_{benchmark}$ of, say, $100/bbl; and,

- Therefore, Market Price $_{newoil}$ =$100-2/bbl=$98/bbl.

This will not give an exact answer because there will numerous other factors that must be taken into account. However the GPW differential provides a starting point for discussions about the sales price of the first few cargoes while the market is getting to know the new grade and the value it is capable of yielding in each market.

The GPW calculation can only be generic because each refinery has different equipment and some are more efficient than others. So each refiner will calculate its own GPW.

From the perspective of a crude oil refiner the quality of the oil is of considerable importance in establishing what it is prepared to pay for a cargo. But refiners are even more interested in the price difference between what it must pay for its feedstock, i.e. crude oil or semi-finished products, and the price at which it will be able to sell the refined products that its refinery plant will derive from the feedstock. This so-called "crack spread" is a key component of the refiner's margin.

## Location and Freight

When it comes to assessing what the market will pay for a particular grade of oil, quality is not the whole story. The location of the oil field or terminal and the freight cost and time taken to deliver the crude feedstock to the refinery gate is a further significant factor in the refiner's margin calculation, as demonstrated in Figure 25.

---

[47] By definition the market price of a benchmark grade is transparent and widely reported.

**Figure 25 Refinery Choices**

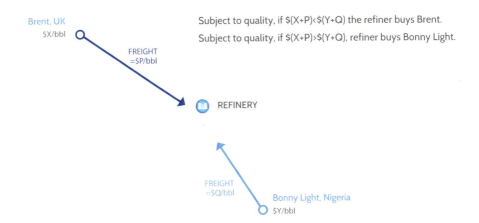

Brent, UK
$X/bbl

FREIGHT
=$P/bbl

Subject to quality, if $(X+P)<$(Y+Q)$ the refiner buys Brent.
Subject to quality, if $(X+P)>$(Y+Q)$, refiner buys Bonny Light.

REFINERY

FREIGHT
=$Q/bbl

Bonny Light, Nigeria
$Y/bbl

Broadly speaking the more flexibility that a loading terminal can offer buyers the more those buyers will be prepared to pay for the oil. Larger vessels tend to have a lower per unit freight cost so the larger the tanker that the terminal can accommodate the greater the price producers will be able to charge for the crude oil. This has implications for the storage capacity and the number and size of jetties that are required by the field design.

As discussed in Chapter Two, this fact is frequently over-looked when the field economics are being constructed prior to the finalisation of a field development plan. Low cost development options may be chosen by joint venture producers because the capital expenditure (CAPEX) needed to install extra storage or more jetties has a more visible impact on the Net Present Value (NPV) of the field than does the on-going under-performance of price over the life of the field. Up-front CAPEX is certainly much easier to model than an on-going price opportunity cost.

Low-cost development options may also have an impact on on-going operating expenditure (OPEX) over the life of the field. For example, if storage limitations mean that vessels regularly have to load oil that has not had the opportunity to settle in storage, the water content of the cargo may be higher than accepted norms, for which the buyer will expect compensation. Or if vessels are frequently delayed awaiting cargo the terminal will receive regular demurrage claims.

For a small oil field with the low production rates and/or constrained storage capacity a price discount might be expected. However if the field is located in the vicinity of other fields such that there is regular tanker traffic in the area, the field may establish a niche for itself in providing "top-up" parcels to larger vessels. This can offset any negative price impact of the field's own logistical constraints, so long as the field can accommodate physically those larger vessels. Large vessels leaving adjacent fields with "deadfreight", i.e. unused freight carrying capacity, may pay a premium for a small parcels of, say, 100-200,000bbls, which would otherwise be uneconomic to ship on their own.

## Any Other Relevant Factors

The oil market is a small world and experience of dealing with a particular terminal, particularly negative experiences, gets around very quickly. For example, if a terminal has a poor record of forecasting production and vessels are regularly delayed awaiting cargo, buyers will pay less for the crude.

If the terminal is frequently shut in by bad weather this too may be reflected in the price. If vessels are regularly held up awaiting customs clearance or other documentation, buyers will make a provision for any unrecoverable demurrage in the price of the cargo. For example, buyers would not buy a cargo from the Black Sea without making provision in their economics for days of waiting to transit the Bosporus Strait.

If terminal metering is inaccurate such that there are regular quantity claims, traders will also build this in to the price they are prepared to pay.

For a seller the "correct" grade differential, G, is the differential at which the best buyer is prepared to buy. If that is influenced by factors that the seller does not consider to be relevant then the buyer has the option of buying an alternative grade from someone else. When the market is tight and oil is in short supply these factors will cease to be relevant and buyers will pay up just to secure the barrels.

## The Time Differential within the Grade Differential - the Missing Link

This sub-section of the book will probably be the most challenging for the non-trader. Crack this section and you are well on your way to understanding how traders think.

Earlier in the book we have discussed the concept of the three elements of the oil price:

- the absolute price, A, which is the price of a benchmark grade of oil such as Brent, Dubai or WTI for delivery in a forward contract month;

- the time differential, T, which expresses the differential between the price of oil for delivery in the time period relevant to the forward delivery month of the benchmark, A, and the actual delivery date of the cargo in question, T;

- the grade differential, G, which expresses the differential between the price of the benchmark grade of oil, A, and the actual grade of oil being delivered under any individual contract.

These three components of the price are built on the premise that the "correct" price of any given spot cargo is the average of published quotations on the 3-5 days around the anticipated bill of lading (B/L) date, or alternatively, depending on the region in question, the average of published quotations on the 3-5 days after the anticipated B/L date.

In the case of term contracts that cover deliveries of a number of cargoes over time when the date range in which each individual cargo will be delivered is not yet known, the three components of the price are built on the premise that the "correct" price of any given cargo is the average of published quotations during the whole month in which the cargo is delivered.

What would happen if, for some reason, the seller of a cargo with an anticipated delivery date range of, say, 15th-17th Month (M) insisted that the price be set in accordance with the average of published quotations on the 3-5 days around the 25th-27th M date range instead of the correct 15th-17th M date range? The seller might want this because of an expected OPEC meeting on the 20th M, which it expected would deliver an increase in the value of the absolute price, A. For any buyer with trading capability this would be a nuisance, but would present no real difficulty.

Holding fast to the principle that the correct price of the cargo is the average of published quotations on the 3-5 days around the 15th-17th M date range, the buyer would simply adjust the price formula by the difference, as evidenced by the CFD market, between the value of cargoes for delivery in the 15th-17th M date range and the value of cargoes for delivery in the 25th-27th M date range.

Let's say, for example, that on the day the deal was struck the market was in contango such that the value of the 15th-17th M date range was $0.10/bbl below the value of 25th-27th M date range. By insisting that the price formula reflects the 25th-27th M date range, rather than the correct 15th-17th M date range, the seller would be over-valuing the cargo by $0.10/bbl. To compensate for this fact the buyer would reduce the apparent grade differential, G, by $0.10/bbl. This would restore the correct value of the cargo at the time the deal was struck and the buyer would then go on to hedge the 25th-27th M date range in the normal way.

As between the buyer and the seller of the physical contract the invoice price of the cargo will now be totally different from what the price would have been if the parties had agreed a price formula that valued the oil on the 3-5 days around the 15th-17th M date range: in between the 15th-17th M and the 25th-27th M anything could happen to the absolute price, A, to deliver an entirely different price result. But the simple act of deducting $0.10/bbl from the apparent grade differential, G, at the time the deal was struck means that the buyer bought the cargo at its fair value at the time it did the deal. So long as the buyer has the capability of hedging the value of A, it is not disadvantaged by giving the seller the price formula it wanted. It may even be able to negotiate a further fews cents/bbl of a discount for accomodating the seller's preference.

From outside of the deal looking in, what is reported is the net grade differential, G, which in this case is comprised of two components:

- The actual grade differential, G; and,

- A -$0.10/bbl adjustment factor to reflect the fact that the price is set by reference to the more valuable 25th-27th M date range, rather than the correct date range of 15th-17th M.

Unless outside observers are privy to all the terms of the deal it may be considered that the grade differential, G, for the grade of oil in question has fallen by $0.10/bbl. In fact this would be totally untrue.

Let the buyer, seller and price reporter beware! The apparent grade differential may be a compound of true grade differential, plus an element of disguised time differential.

This Chapter has completed our look at the three components of the oil price, the absolute price, A, the time differential, T, and the grade differential, G. Chapter Six will build on this knowledge by examining in more detail what can be done to evaluate the price of crude oil and to manage oil price risk. It will conclude our exploration of the trading instruments used by market participants for both speculative and hedging purposes, with a consideration of the swaps and options markets. Options are traded both on regulated exchanges and in the OTC market.

# CHAPTER 6

❝ Risk comes from not
knowing what you are doing ❞

*Warren Buffett*

## Hedging

Hedging is an activity undertaken by companies attempting to take control of their own cash flow by ironing out price spikes and troughs in their oil acquisition or sales contracts. Its objective is to reduce future oil price uncertainty and may be seen as having a risk reducing motivation. Hedging is often necessary because of the way in which oil contracts are priced.

For example, it may be thought that if a refining company wants to ensure that it never pays more than, say, $120/bbl for its oil next year, it should buy the physical cargoes it will need next year when the price of oil is below $120/bbl. Unfortunately the oil market does not work like that for most grades of oil, except for the benchmark grades mentioned in Chapter Three.

A refiner might be able to buy and *take delivery today* of all the oil it will need next year at $120/bbl, depending on the current level of the forward curve. But the cost of capital and the storage costs would be crippling. Oil delivered today for use next year has to be paid for today and stored somewhere until next year when it is needed. This is an impractical solution for most companies.

A buyer of oil may well be able to enter into a term contract that guarantees it a secure supply for next year. But, as we have identified in previous Chapters, the price that it will pay for this supply will typically not be fixed today at a price expressed as $X/bbl fixed and flat. The price formula in a physical purchase contract will refer to the price of oil as it will be in the month or quarter in which each individual cargo will be delivered.

So in order to lock in the price of crude oil today for delivery next year or in subsequent years, oil buyers and sellers must enter into a hedge using one of the financial instruments that trade on the basis of fixed and flat prices rather than floating formula prices. These are forwards, futures, swaps and options contracts.

In Chapter Three we described the use of forward contracts and futures contracts to hedge the absolute price A. In Chapter Four we described the use of a very particular swap, the Dated-to-Paper CFD that is used to hedge the time differential, T. We will now complete the set by describing the use of over-the-counter (OTC) swaps and options and how they too are used to hedge the value of the absolute price, A.

Typically forward and futures contracts are used to implement operational hedging decisions on individual cargoes in the short to medium term. Swaps and options tend to be used to implement more strategic long-term hedging decisions but this is not a universal rule and there are plenty of exceptions.

## Strategic and Operational Hedging Compared

For the purposes of this book we will define a strategic hedge as one that seeks to secure a company's development strategy. If a company's future is linked to the price of crude oil it may wish to take control of its price risk with a strategic hedge. This may involve under-writing the base case price assumption in a corporate budget. Or it may involve guaranteeing a price, or profit margin, that will secure the economics of a particular asset or project. In some cases a strategic hedge may be undertaken at the insistence of a bank that is providing loan finance to develop a project.

Typically a strategic hedge is one that is undertaken on a significant scale relative to the size of the company and one that will have an impact on the fortunes of the company in the medium to long term.

Strategic hedging differs from the operational hedging that we have described so far in this book. By operational hedging we mean guaranteeing the profit margin or locking in the price of a particular cargo. Operational hedging is typically a smaller scale activity taking place over a shorter time horizon.

Consequently the tools that are used for strategic hedging are somewhat different from those used for operational hedging. Broadly speaking, and with many exceptions, forward and futures contracts are associated with operational hedging while OTC swaps and options are associated with strategic hedging.

## Brent Again

As mentioned in Chapter Three what is commonly referred to as Brent is the most pervasive of all the benchmark grades currently used as a price reference point in an estimated two thirds of the world's physical oil contracts. It is also the grade that underlies the Intercontinental Exchange's (ICE's) flagship futures contract. The Chicago Mercantile Exchange (CME) also offers a Brent futures contract, although the jewel in its crown is the WTI contract.

The OTC swaps and options market tailors contracts to the needs of oil buyers and sellers. Data on OTC contracts is sparse, which is one of the reasons that regulators are so concerned about the market, so it is impossible to say what proportion of oil swaps and options are expressed in terms of Brent and how many are expressed in terms of WTI or Dubai or ASCI. But since a large proportion of physical buyers' and sellers' oil price risk is based on Brent it stands to reason that a large proportion of OTC oil swaps and options must be similarly expressed in terms of Brent.

There is a virtuous, or some might say, vicious, circle at work here. The more physical contracts that are based on Brent the greater the volume of trade in risk management tools that is also based on Brent. The greater the liquidity (i.e. volume of trade) and the greater the flexibility in the range of risk management tools on offer the more likely it is that financial actors, i.e. speculators, hedge funds etc., will be attracted to the market, particularly if the underlying price is a volatile one that offers a profit opportunity. Also, the more flexible and liquid are the risk management tools expressed in terms of Brent, the more producers and end-users are likely to use Brent as a reference point in physical contract price formulae.

Table 5 describes the range of Brent-based contracts that is available for physical trading, risk management and speculation and the uses to which each is put.

**Table 5 The Brent Contracts**

| | Wet Brent or Dated Brent | 25-Day Forward BFOE | Brent CFDs or Dated to Paper Swaps | Brent ICE Futures (see also CME) | Brent OTC Swaps and Options |
|---|---|---|---|---|---|
| Physical Delivery P Cash Settlement C | P | P or C | C | C, but with the possibility of "delivery" by EFP | C |
| Normal Trading Time Horizon | 10-25 days forward. In theory may trade up to 55 days forward but in practice contracts for more than 25 days forward are treated as 25-Day BFOE. | From 25 days to about 6 months forward. | From 1 week to about 3 months forward | From 25 days to 12 months forward and 4 quarters thereafter. | From 3 months to about 5 years forward, although can be further forward. |
| Trading Unit Size | 600,000 bbls +/- 5%. | 600,000 bbls +/- 1%. | > 50,000 bbls. Upper limit subject to credit. | > or = 1,000 bbls. Upper limit subject to credit. | > or = 100,000bbls. Upper limit subject to credit. |
| Common Usage | Disposal of physical production and purchase of refinery supply. Some limited speculative trading. Sets the value of (A), (T) and (G) usually by reference to a formula of average published price around the bill of lading date. | Hedging and speculation on the value of (A), or spreads between months, the value of (T). Some disposal of physical production/ and buying of refinery supply. | Short-term hedging and some speculation on the value of (T). | Short-term Operational hedging and speculation. Typically involves the value of (A) or spreads between months, the value of (T). | Mainly long-term strategic hedging and project financing. Also speculation. Usually involves the value of (A). |

Re-visiting the earlier hypothetical example, referred to at the beginning of this Chapter, of a refining company that wants to ensure that it never pays more than $120/bbl for its oil next year the refiner will be obliged to enter into a financial derivative instrument, or "paper" contract, that seeks to manage the price formula to which it is exposed under its physical purchase contract.

A "paper" contract is a broad term referring to some form of oil derivative instrument. This may be a forward, futures, or OTC swaps or options contract. Typically "paper" refers to the fact that physical delivery is neither desirable nor, in most cases, possible[48].

The hedger may begin its hedge by choosing the price at which it wishes to limit its risk, say $120/bbl, and locking this price in with a derivative instrument. This can only be done if the market presents the opportunity to do so at levels that the hedger finds acceptable. The mechanics of hedging usually mean that:

- the physical oil *buyer* starts by purchasing a fixed price paper derivative contract. This paper contract is subsequently cashed-out by selling the paper contract at a formula price. In a perfect hedge the formula purchase price of the physical cargo is exactly offset by the sales price formula of the derivative contract.

- the physical oil *seller* starts by selling a fixed price paper derivative contract. This paper contract is subsequently cashed-out by buying the paper contract at a formula price. In a perfect hedge the formula sales price of the physical cargo is exactly offset by the purchase price formula of the derivative contract.

## Important Point about Hedging

Hedging does not necessarily minimise price for a buyer, nor maximise the oil price for a producer. The overall result of a physical purchase or sale and a hedge is to lock in the price at which the hedger decides to place its hedge:

- If a refiner decides to hedge and the market price subsequently falls it may wish that it had not hedged because it would have achieved a lower price from the purchase formula under its physical contract.

- If a producer decides to hedge and the market price subsequently rises it may wish that it had not hedged because it would have achieved a higher price from the sales formula under its physical contract.

---

[48] In the case of the oil market the precise terminology should be treated with care. Some of the most frequently traded oil 'paper' hedging and trading instruments are the 25-Day Brent/Forties/Oseberg/Ekofisk ('BFOE') forward contract and the Dubai swaps contract. The former is a contract where the seller has the option of delivering one of the range of grades of crude oil specified. In both cases the contract is based on oil that may be delivered physically, but in most cases involves oil trades that are settled in cash.

A loss on a paper hedge transaction is more visible than an offsetting gain on a physical cargo contract and therefore can often become the main focus of attention. A loss on the paper side of a hedge transaction is often referred to as a "hedge loss", although this term is inaccurate. If a hedge is properly constructed for every loss there is an offsetting gain. The "glass half-empty" way of looking at it is that a hedger has, by definition, to make a loss on either the paper derivative contract or on its physical transaction[49]. Or, the "glass half-full" way of looking at it is that the hedger has to make a gain on either the paper derivative contract or on its physical transaction.

A real hedge loss would occur if the buyer had tried to lock in a net overall price of $120/bbl, but in reality ended up with an overall result from the hedge and physical transaction of, say, $115/bbl or $125/bbl. This might occur if the hedge had not been properly constructed and the paper contract purchase price formula did not match the physical cargo sales price formula.

## Correlation and Basis Risk

A perfect hedge is one where for every 1 cent/bbl the paper hedge price rises, the physical price also rises by 1 cent/bbl and where for every 1 cent/bbl the paper hedge price falls, the physical price also falls by 1 cent/bbl. In other words the ideal situation is where there is a 100% positive correlation between the price of the physical contract and the price of the paper hedge contract.

There is only a handful of forward benchmark crude oil contracts that trade at fixed prices and which therefore lend themselves for use as hedging instruments. However there are hundreds of grades of crude oil that trade under physical contracts at formula prices that need to be hedged. Hence the correlation in price between the forward grades and the physical grades is unlikely to be exactly 100% because the value of the grade differential, G, does not rise and fall cent for cent with the absolute price, A.

The lower the correlation between the physical price index and the paper hedge price index, the less perfect the hedge. A poor correlation gives rise to "basis risk", i.e. hedging the price of apples with a paper contract in oranges.

Similarly timing differences, i.e. differences in the value of the time differential, T, can produce basis risk in oil contracts. For example, say a producer sells its oil on physical contracts with a formula price related to a published price average related to the bill

---

[49] The only time this would not be true is if prices were exactly the same at the time the hedge is placed and the time when the physical transaction is undertaken and the paper hedge contract is cash-settled.

of lading (B/L) date, but hedges this price with a futures contract. When it comes to sell its physical oil the producer will achieve the price for "Dated" oil on the 3-5 days around or just after the B/L date. It will close its futures hedge by buying oil futures for delivery probably 1-2 months forward. It will purchase its futures contracts to close its hedge position over the 3-5 days around or just after the B/L date. But the oil it will be purchasing on those dates will be oil for delivery 1-2 months forward, not wet, physical Dated oil. In other words it will be hedging the price of apples for delivery in the next 25 days with a paper contract in oranges (G basis risk) for delivery in 1-2 months' time (T basis risk). We discussed this in Chapter Three.

The time differential basis risk can be managed by grafting onto the hedge of the absolute price, A, a further CFD hedge of the time differential, T, as discussed in Chapter Three. Grade differential basis risk cannot be managed by hedging. It can only be minimised by choosing a benchmark reference price in the physical oil price formula that has a good correlation with the paper contract price that will be used to hedge it.

Ensuring a good correlation between the physical oil price risk and the hedge instrument used to hedge it is nowhere more important than in the upstream oil industry, as discussed in the next section of this Chapter.

## Tax Basis Risk

In Chapter Two of this book we looked at the operation of Production Sharing Contracts (PSCs) and noted that when considering the profitability of an individual oil field, there may be an effective "ring-fence" applied to ensure that oil sales from a field are used to recover only those costs that are directly attributable to the exploration for and development of that field.

Taxation inside an oil field's ring fence (IRF) may very well be at a different rate to the corporation tax rate applied outside the field's ring-fence (ORF). Typically oil production is taxed at the IRF tax rate and any hedges undertaken by the producing company are taxed at, or attract tax relief at, the ORF tax rate. Hedge "losses", for example, cannot be claimed as recoverable costs IRF.

A similar issue can arise in a multi-national refining company in the situation where the refinery operates in one country, but the hedges of its refining margin are undertaken by a different affiliate in a different country. Each may attract a different rate of corporation tax. However a refiner may rectify imbalances in the tax treatment of different aspects of its operations by re-organising or relocating its trading activity to achieve hedge accounting treatment of transactions undertaken to hedge its physical purchases and sales of crude oil and refined products.

For a producing company the disparity between IRF and ORF tax rates has to be managed, particularly since *effective* IRF tax rates can be up to 80%, whereas corporation tax rates may be as low as 20%. Consequently if a producing company gains because its physical oil sales price rises with the market, it will lose 80% of this rise to the tax man. By definition in these circumstances the oil producing hedger must lose on the paper hedge side of its equation. But it will only receive tax relief for these losses at the much lower corporation tax level of 20%.

To illustrate the issue let's assume that a producing company (Producer) is taxed IRF at, say, 70% on its production of Generic Blend oil. Let's say that ORF the Producer is liable to Corporation Tax at a rate of 30% and can get tax relief for losses at the same rate. So if it hedges Generic Blend using the Brent Market, or Dubai swaps or WTI its gains or losses on the hedge are taxed, or are allowed tax relief, at 30%.

Let's now assume that in December the Producer decides to hedge a cargo of 600,000 bbls of Generic Blend for delivery in the following March using the Brent market. In December it sells forward a cargo of 600,000 bbls of Brent for March delivery at, say, $100/bbl. On or about mid-February the Producer is able to sell its cargo of Generic Blend because the terminal operator has issued the schedule of loadings for March, showing the Producer the delivery date of its cargo of Generic Blend. The price has fallen to $90/bbl, so the Producer sells its cargo of Generic Blend at $90/bbl.

The Producer sold forward a cargo of Brent Blend for March delivery in December at $100/bbl. It now has to prepare to deliver a Brent cargo in March. It does not have one so it must go into the market and buy a cargo of March-delivery Brent at $90/bbl[50]. Hence it makes a $10/bbl profit on its hedge. So the Producer has sold Generic Blend at $90/bbl and has a $10/bbl profit from its hedge to add to this, giving it a net sales price of $100/bbl. This is the price the Producer decided it would like to lock in back in December when it entered into the Brent hedge, so the hedge has worked perfectly. Now let's look at the after tax result.

---

[50] To illustrate the tax issue we are assuming a simplified hedge with none of the other practical complications that we discussed in Chapter Three.

**Table 6 The Impact of the Differential Tax Rate-Hedge Profit**

|  | Generic Blend | Brent | Total |
|---|---|---|---|
| Sales Price ($/bbl) | 90 | 100 |  |
| Purchase Price ($/bbl) |  | 90 |  |
| Revenue ($/bbl) | 90 | 10 |  |
| Cargo size (bbls) | 600,000 | 600,000 |  |
| Revenue ($) | 54,000,000 | 6,000,000 | 60,000,000 |
| Tax Rate (%) | 70 | 30 |  |
| After Tax Revenue ($) | 16,200,000 | 4,200,000 | 20,400,000 |

Table 6 shows the impact of the different tax rates IRF and ORF. The total after tax revenue from the physical cargo sale and from the hedge taken together is $20,400,000.

To generate that amount of after tax revenue from a physical sale alone, the Producer would have had to sell its Generic Blend cargo at $113.33/bbl, i.e. ((Revenue/ Percentage of Retained Income)/ the number of barrels) = (($20,400,000/30%)/ 600,000bbls) =$113.33/bbl. But for a hedge to be perfect the Producer should receive the same after tax revenue as it would have received if it had sold the Generic Blend cargo at $100/bbl, so this hedge does not work. In this case the Producer has enjoyed a windfall gain from the effect of the differential tax rate, but if the oil price had gone up instead of down the picture would have looked very different as shown in Table 7.

**Table 7 The Impact of the Differential Tax Rate- Hedge Loss**

|  | Generic Blend | Brent | Total |
|---|---|---|---|
| Sales Price ($/bbl) | 110 | 100 |  |
| Purchase Price ($/bbl) |  | 110 |  |
| Revenue ($/bbl) | 110 | -10 |  |
| Cargo size (bbls) | 600,000 | 600,000 |  |
| Revenue ($) | 66,000,000 | -6,000,000 | 60,000,000 |
| Tax Rate (%) | 70 | 30 |  |
| After Tax Revenue ($) | 19,800,000 | -4,200,000 | 15,600,000 |

The total after tax revenue from the physical cargo sale and from the hedge taken together is $15,600,000.

To generate that amount of after tax revenue from a physical sale alone, the Producer would have had to sell its Generic Blend cargo at $86.66/bbl, i.e. ((Revenue/ Percentage of Retained Income)/ the number of barrels) = (($15,600,000/30%)/ 600,000bbls)

=$86.66/bbl. This falls far short of the after tax revenue it would have achieved if it had sold the Generic Blend cargo at $100/bbl.

The quick fix for this disparity is to scale down the hedges by the ratio of retained income IRF to retained income ORF.

For example, let's assume that the IRF oil production tax rate is 70% for a particular field and the ORF corporation tax rate for the company is 30%. The scaling factor that must be applied to balance up after-tax gains and losses across the ring fence is:

$$(100-70)/(100-30)=0.429$$

Hence to hedge a cargo of 600,000 bbls, requires a hedge volume of only 257,143 bbls, i.e. 42.9% of 600,000 bbls. This would require hedging with 257 lots of futures, rather than a 600,000 bbl cargo of 25-Day BFOE.

As shown in Table 8, scaling down the hedge by this factor of 0.429, which applies only to the 70% IRF: 30% ORF case, delivers the same after tax revenue that the Producer would have achieved if it had sold the Generic Blend cargo at $100/bbl i.e. ((Revenue/ Percentage of Retained Income)/ the number of barrels) = (($18,000,000/30%)/ 600,000bbls) =$100/bbl.

**Table 8 The Impact of the Differential Tax Rate- With a Scaled Hedge**

|  | Generic Blend | Brent | Total |
|---|---|---|---|
| Sales Price ($/bbl) | 110 | 100 | |
| Purchase Price ($/bbl) | | 110 | |
| Revenue ($/bbl) | 110 | -10 | |
| Cargo size (bbls) | 600,000 | 257,143 | |
| Revenue ($) | 66,000,000 | -2,571,429 | 63,428,571 |
| Tax Rate (%) | 70 | 30 | |
| After Tax Revenue ($) | 19,800,000 | -1,800,000 | 18,000,000 |

This scaling factor has to be kept under constant review particularly with a new field where costs are still being recovered. Once cost recovery pay-back is reached the state's profit share may start to apply causing a sudden shift in the effective IRF tax rate. It can be tricky to calculate in advance the point of payback as this will depend on the absolute price of oil that the state in question uses to calculate cost recovery. The higher the oil price, the quicker cost recovery will be achieved[51].

This can give the producer volume uncertainty in its hedges, which are by definition undertaken when the price of oil for delivery in a particular year is no more than a forecast. Consequently it may be unwise for a producer to hedge 100% of its after tax exposure on an oil field where the tax rate may be subject to a quantum leap.

Alternatively the producer may prefer to undertake its hedges using an options strategy where the consequences of being over-hedged are easier to quantify and to limit. We will discuss hedging with options later in this Chapter.

As mentioned above, in the case of operational hedging oil price hedging is usually undertaken around 2-6 months in advance of the delivery date of the physical oil being hedged. In the case of strategic hedging the hedge may be undertaken from 3- 6 months up to several years in advance of the delivery date of the physical contract being hedged. It will be recalled from Chapter Two that the physical delivery date of individual cargoes does not emerge until the terminal or field operator's schedule of liftings is issued for month M, around about 5-10th M-1, or in some regions considerably earlier.

Also as mentioned above, in the case of operational hedges the instrument of choice tends, with exceptions, to be either forward or futures contracts. In the case of strategic hedges the instruments chosen are usually the OTC swaps and options market.

# Horses for Courses - Choosing the Right Tool for the Job

It will be recalled from Chapter Three that the forward market is labour intensive involving physical cargo nominations and liftings and the provision of a performance guarantee in the form of a letter of credit or letter of indemnity. The futures market is considerably less labour intensive than the forward market, but involves the upfront payment of initial margin and a daily transfer of variation margin. Variation margin must be paid when a hedge position is losing money, irrespective of the fact that the physical oil it is hedging is making an offsetting gain. This means that neither the forward nor the futures markets is ideal for use by the strategic hedger in large scale, long term hedge positions.

The OTC swaps and options market can be more attractive to the oil company and refining company hedgers because the hedge does not require the physical transfer of

---

[51] Consilience has designed an application that permits the tax department and the trading department of producing companies to explore together the impact of its tax, expenditure and price assumptions on its hedge ratios. This can be made available to readers on request.

oil and all the shipping, volume, measurement and documentation risk that accompanies such a physical transfer. Furthermore, the need to provide performance security is minimised by the fact that the derivative provider is able to take the hedger's offsetting physical position into account when establishing credit lines with the hedger.

As mentioned above OTC derivatives can be divided into swaps and options. We will consider both during the balance of this Chapter. But before we do there are some general principles that a company must take into account before embarking on a trading or hedging strategy.

In planning its strategy a company usually starts by considering:

- its risk profile;

- its risk appetite; and,

- its market price view.

We will consider each of these in turn.

# Risk Profile

Companies that welcome oil price risk and for whom risk-taking is a core business activity are adept at modelling, measuring and managing their own price risk. Trading companies and banks have specialised personnel whose sole job it is to understand how the value of the company's portfolio changes with every cent that the oil price rises or falls.

Industrial oil firms, meaning independent oil producers and refiners, typically have a corporate model into which a price base case assumption is plugged and tested with high and low price case sensitivities. In most companies any decision to manage oil price risk is typically taken centrally perhaps by the board or by a risk committee. In other cases subsidiary companies or asset teams or project managers have authority to manage their own "departmental" price risk.

No sensible price risk management decisions can be taken by a company without a clear idea of its own risk profile at varying price levels, so this has to be the starting point for any hedging strategy. This risk profile is rarely linear.

The corporate plan of an exploration and production (E&P) company is typically framed in terms of alternative high, low and base case price scenarios, with progressively more projects capable of being undertaken or with more exploration wells capable of being drilled at higher price levels. Oil price is, of course, only one part of the story. A bad project is a bad project regardless of the price. In a well-managed E&P company funds are only allocated to projects or assets that pass the company's hurdle rates of return, regardless of how much money is in the corporate war chest. Nevertheless E&P companies can plan to do more at higher oil price assumptions than they can do at lower prices.

E&P companies rarely plan based on a high price scenario and will organise their activities typically using their base case assumption. Even so, they will often hold cash in reserve in case the low price scenario comes to pass.

When times get tough there is some limited flexibility to delay wells or shelve development plans, but most E&P developments take place as joint venture operations. So a decision to change the spending plan in a field budget after it has been agreed requires the cooperation of the joint venture partners with whom the budget has been agreed. If that cooperation is not forthcoming, and consequently an E&P company cannot meet its cash calls from the operator in accordance with the approved joint venture operating committee budget because the oil price turns out to be lower than expected, the company runs the risk of losing its share of the field. Discretionary spend on staff and central services can be cut when oil prices are low, but these items tend to be small compared with the exploration and development budgets.

An E&P company's high price planning sensitivity would allow it to develop a wider range of assets and drill a greater number of exploration wells than would be possible under the base case price sensitivity. So in theory it may be expected that, if the opportunity arises to lock in by hedging an E&P company's high price case, this should be a "no-brainer" for it. But in practice this is not the case as discussed later in this Chapter.

A refining company has a slightly different decision-making process because refineries typically have a high level of fixed and sunk costs. The hedging process therefore is more focussed on operating costs and the profit margin between the price of buying crude and the price at which refined products can be sold, the so-called "crack spread".

A positive refining margin, however small, makes a contribution to sunk costs so refineries will plan to run even when the margin is low. Refiners can hedge their crude oil purchases comparatively easily with the same grade and time differential risks faced by producers on the other side of the equation. But the tools to hedge product sales are blunt instruments at best.

There are hundreds of different grades of oil products sold in the market worldwide and the specification of these grades changes from region to region and from season to season. The trading instruments available to hedge these products are considerably fewer. Therefore the oil product price hedger faces a much higher level of basis risk. The hedging of refined products is considered in the companion volume to this book "Trading Refined Products: The Consilience Guide[52].

## Risk Appetite

Risk appetite describes the willingness of a company to accept risk arising out of oil price volatility. This is a function of the core business of the company and its culture. Trading companies are in the business of taking measured price risks, but for many industrial firms price risk is something they would prefer not to have to address.

For example, the business of an E&P company is to find oil and bring it on-stream successfully. Such companies may enter into high risk exploration plays at which much larger companies would baulk, because this is the E&P company's core business activity to which its skill set is tailored. However such companies often have no confidence in their own ability to cope with oil price risk once they have found oil. They do not regard this as their core strength.

Such companies typically sell their physical oil to a larger company under long term contracts, usually annual contracts, at published floating market average prices. They sometimes may be forced to enter into hedges of a substantial proportion of this floating market average oil price risk at the insistence of their bank as a condition of lending to them to finance the development of the project.

Some other companies prefer to hedge none of their oil price risk. There are three reasons frequently cited for not hedging:

• First, the belief that oil company shareholders invest in these shares because they want to take on oil price risk. It can be pointed out that this argument does not quite hold water. If shareholders really seek oil price risk then buying shares in oil companies is a peculiar way of satisfying that need. The correlation between oil share prices and oil prices themselves is quite low. Oil share prices reflect the quality of the management team and the market's view of the prospectivity of its licensed exploration acreage. Nevertheless the opinion that shareholders want oil price exposure is deeply entrenched in the psyche of E&P company executives because it is a view often expressed to them by their shareholders, rightly or wrongly.

- Secondly, the belief that entering into a hedge implies that the company has an oil price view and is backing that view by "trading" in the market. Oil company executives are quick to point out the volatile nature of oil prices and the fact that they are fundamentally unforecastable other than in a very broad range. Going into the market to hedge at, say, $100/bbl may be taken by shareholders to imply that the producing company believes that the price will not go above $100/bbl or that the consuming company believes that prices will not fall below $100/bbl. Examples are legion of companies that hedged and locked their shareholders out of subsequent favourable price moves being vilified in the press and suffering a fall in their share price. Conversely companies that have oil price exposure, but do not hedge, may be accused of having an implicit price view that the market will move in their favour. However, if the price moves against the unhedged company, shareholders tend to be more phlegmatic because oil companies are not responsible for market price movements.

- Thirdly, in many instances of high oil prices there is significant backwardation in the forward oil price curve and at low oil prices there is significant contango in the oil curve. This is not always the case, as described in Chapter Four. But there is a tendency for the forward oil price curve to exhibit backwardation, i.e. a positive time differential, T, when the absolute level of the oil price, A, is at a high level. Similarly there is a tendency for the forward oil price curve to exhibit contango, a negative time differential, T, when the absolute level of the oil price, A, is at a low level. Producers attracted by a high absolute price at the front end of the forward price curve, i.e. by the price of oil for "prompt", or immediate, delivery, are often disappointed when they try to implement a strategic hedge up to two years forward. This is because they have to sell at a much lower level than the price of prompt delivery oil may have led them to believe. Similarly, oil consumers attracted by a low absolute price at the front end of the forward price curve are equally disappointed to find that if they try to implement a strategic hedge up to two years forward they will be buying at a much higher level than the price of prompt oil may have led them to believe. As discussed in Chapter Four, the hedger may hedge the slope of the curve and the height of the curve as two separate and independent hedges. But for most industrial firms this may well be a hedge too far.

Consequently unhedged oil companies are some of the biggest oil price risk takers in the market. Deciding not to hedge forces the company to hold capital in reserve to cope with volatility in its cash flow; this is capital that might be more efficiently deployed in growing the business. Not hedging carries the risk that the company will have to change its corporate strategy in mid-plan to respond to adverse price moves and potentially downsize the business. Nevertheless a lot of companies, particularly producing companies, do not hedge. You can lead a horse to water, but you can't teach it synchronised swimming!

To hedge successfully a company would be well advised to clarify its risk appetite upfront and explain to its shareholders why it is undertaking hedges, if it chooses to do so, and what the consequences could be of hedging and of not hedging. If a company does not hedge and the price moves against its physical position, it is facing an opportunity cost: if it had hedged it would have locked in more favourable prices for its physical oil. If a company does hedge and the price moves in favour of its physical position, it will have to pay out cash to the hedge provider: if it had not hedged it would not have a loss on the paper side of the equation.

Many oil firms treat the two possible outcomes unequally. An opportunity cost is treated less negatively than a paper "loss" because the latter will have to be quantified and be exposed in the company accounts.

Companies that hedge and then lose on the paper side of the equation, as opposed to suffering an opportunity cost on the physical side of the equation, often act as if they have had their fingers burnt and shy away from hedging in the next planning cycle. Given the cyclical nature of pricing that may be the very time when they should hedge again because the next price move arguably stands a higher chance of being in the opposite direction.

## Market Price View

A company's market price view refers to the company's opinion as to whether the price of oil is more likely to rise or to fall over the time horizon under consideration. Companies exposed to the price of oil have to have a view on likely future price movements, usually spelled out in its scenario plan as a high, low and base case assumption, as described above.

Once companies have modelled their price risk profile, considered their appetite for risk and devised their strategy for coping with actual price outcomes at different levels they may decide to hedge all or part of this risk. It is an unavoidable consequence of the decision to hedge that the company must begin to assign rough probabilities to its various possible price scenarios. This is because in implementing a hedge strategy the company must decide what outcome it is actually hedging against in order to start placing hedges at particular price levels when the market presents the opportunity to do so. In other words the hedger faces the decision of at what price it will enter the market to hedge.

There is a very human tendency in industrial oil firms, as opposed to trading firms, to consider that the right time to start hedging is when the market is, say, $2/bbl better than it is now, whatever that price level may be right now. As the market moves towards that target there is a tendency to move the target too. If on the other hand the market moves further away from the target there is often a sense of frustration about the missed opportunity. This may render the company impotent watching the market move further and further away from its initial target, particularly in organisations that have a "blame" culture. It is sometimes the case that, when it finally dawns on the frustrated hedger that the next oil price move might threaten the future of the company, hedges are at last put in place at less than ideal oil price levels.

There are many examples of producing companies hedging at very low prices and consuming companies hedging at very high prices, then being crucified by shareholders for having hedged just before the market turned back in favour of the company's underlying physical position.

Hedging requires a level of discipline of which many industrial firms are incapable because oil pricing is not their core business activity. The novice hedger will always attempt to sell at the high of the market, in the case of the producing hedger, or buy at the low of the market, in the case of the consuming hedger. Each is likely to miss the peak or the trough just before the market starts to turn against them. This leads to producers chasing the market down and consumers chasing the market up. For producers it is advisable to scale into a hedge in a rising market and for consumers to scale into a hedge in a falling market. Then if the market turns, at least some volume is already hedged.

## When to Close a Hedge

The other issue with which industrial firms struggle in the hedge decision-making process is when to close strategic hedges. Theorists hold fast to the principle that once a hedge is put in place it should remain in place until the oil it is hedging is priced. The hedge should be closed at exactly the same time as the physical contract price formula is resolved, say the theorists.

For short-term operational hedging this is broadly true. If a hedge locks in the price margin between buying and selling a cargo of oil, lifting the hedge too early puts the profit margin at risk. Trading companies may decide to do this because they want to take on additional price risk, i.e. speculate. But for industrial firms the "right" time to close an operational hedge of a particular cargo is when the value of the price formula under the physical contract is established, probably on 3-5 days around or just after the B/L date of the cargo as discussed in previous Chapters.

In the case of long-term strategic hedges the decision when to remove a hedge is less clear-cut. As discussed in the previous three sections of this Chapter hedges may well be best undertaken after an analysis of a company's risk profile, risk appetite and market price view. If any or all of these three guiding factors change it may be appropriate to re-visit the hedges that are already in place and to decide to close them early.

For example, if the price of production from a new oil field has been hedged, but the field start-up is delayed, it may be appropriate to close the hedges that were put in place to cover initial production. Or it may be that a share of the company's stake in the oil field is sold to a third party, in which case hedges of that share may no longer be appropriate. Similarly, if the company has access to a new source of financing, perhaps from a rights issue, it may be that the company can support its business plan even if oil prices move against its physical position. Again it may want to re-visit its hedges in these circumstances.

Perhaps the most contentious reason for closing hedges early is when a company's market price view changes. If a company has hedged against the risk of oil prices falling below, say, $70/bbl and a war breaks out that shuts in a significant volume of world oil production, the hedger may decide that the risk of a fall below $70/bbl has gone away for the foreseeable future and close its hedges. Or it may be that Iraq succeeds in its objective of establishing spare production capacity of several millionb/d, in which case a consumer may decide that the risk of prices rising above $150/bbl has disappeared and its hedges against that eventuality are no longer appropriate.

Closing hedges early is much easier to do if the hedges are making money when the decision to remove them before they mature is taken. Early closure undoubtedly re-opens oil price risk. So if the hedger does this, and it turns out the revised price view was wrong, there will be questions asked about whether or not the company is actually speculating rather than hedging. These questions will be all the sharper if the hedges have been closed early at a loss.

There is no infallible objective test of whether an activity is hedging or speculating. The same action may be either depending on the risk profile, risk appetite and market price view of the company concerned and its intention at the time it took the action. All that can be said with certainty is that it would be illogical to persist slavishly with hedges that were opened based on circumstances that have subsequently changed.

## When Strategic Hedges Go Operational

Ultimately all strategic hedges become operational hedges over the passage of time. This is when some adjustment to positions may be necessary whether the company's risk profile, risk appetite or market price view have changed or not. This is because when strategic hedges are placed in the market they tend to be associated with broad tranches of oil which have not yet been organised into cargo lots with identifiable lifting dates. Hedges need to be lifted when the oil that they are hedging is priced in the market in cargo lots with associated dates. As discussed in previous Chapters, cargoes tend to be priced in accordance with a formula related to published prices around or just after the bill of lading (B/L) date.

To recap on price formulae, so far in this book we have analysed the price of oil in accordance with the three components that make it up: the absolute price (A), the time differential (T) and the grade differential (G). We have looked at the physical market and noted the price convention that says that the "correct" price formula for a given cargo of oil in the spot market tends to be the average of published quotations on the 3-5 days around the anticipated B/L date, or in some regions on the 3-5 days after the anticipated B/L date.

We have noted that when two traders transact a spot cargo they may identify the three price components referred to above explicitly. For example, the traders may agree that the price formula will be:

- the average of second month 25-Day BFOE on the 5 days after the B/L date; plus/minus

- the value of the time differential as evidenced by the CFD market at the time the deal is struck, i.e. the value in CFD market of the differential between the price of cargoes that have the dates of the cargo in question and the price of 25-Day BFOE; plus/minus

- a fixed grade differential of $X/bbl.

However it is more common for traders to recognise the three components of price implicitly by referring to the price of a specific grade as published on the 3-5 days referred to in the price formula in the physical contract. For example, the traders may agree that the price formula will be the average of Bonny Light prices as published on the 5 days after the B/L date. In this case the values of A, T and G are all left to "float" and be determined by their published values on the 5 relevant days.

It is typically the case that in this latter type of deal the three price components only become explicit when one or other counterparty to the deal decides to hedge the value of A and/or T. (It will be recalled that the value of G is not easily hedgable, except in the case of the grade differential between two benchmark grades and to a limited extent in Urals Blend).

In the case of the example above of the cargo with the price formula of the average of Bonny Light prices as published on the 5 days after the B/L date, let's say the refining company buyer wishes to hedge the value of the absolute price, A. It could do so by buying at a fixed price in the Brent forward market. It will close this hedge by selling 20% of the hedge volume on each of the 5 days after the B/L date when the absolute price of the physical Bonny Light cargo is established by reference to a publication.

If the same refining company also wishes to hedge the value of T it will do so by buying a CFD swap[53] at a fixed price. This CFD hedge will cash settle automatically by reference to the difference in the average values of Dated Brent and futures Brent over the 5 days after the B/L date.

The trading company's only remaining exposure is to the published price differential between Dated Brent and Bonny light over the 5 days after the B/L date.

However, if the refining company starts off by entering into a strategic hedge, say, a year in advance of purchasing a specific cargo, it is more likely to do so using an OTC derivative instrument that hedges only the value of the absolute price, A, for the future month in question. *These OTC instruments are likely to cash settle based on a monthly or quarterly published price average.*

So once the cargo dates become known for any given trading month M, at some time during the first ten days of month M-1, the strategic hedger will have to adjust the strategic hedge to cash settle over the same 3-5 days over which the physical cargo it is hedging is being priced. This is most easily done by grafting on a CFD hedge at that point.

This disparity between the cargo pricing period and the strategic hedge cash settlement period should not be ignored. As we identified in Table 1 in Chapter One the difference

---

[53] The DFL market is directly comparable to the CFD market. It is a market in the differential between the price of Dated Brent and the price of first contract (frontline) futures Brent. It tends to be traded on a fixed price basis and cash-settled based on the average differential between the price of Dated Brent and frontline futures Brent in monthly, quarterly or yearly average periods.

between a three day average price and a monthly average price can be more than $10/bbl.

We will now turn to the instruments that are typically used for the purpose of strategic hedging.

# Swaps

Over-the-counter (OTC) swaps are typically, but not exclusively, used for strategic hedging purposes. Asset or project hedgers may buy or sell OTC swaps in strips of months, quarters, half-years or in annual tranches up to 10 years in the future. However, after 3-4 years forward the market is illiquid and the bid-offer price spread is very wide.

To recap, the term OTC refers to transactions between two named counterparties on terms agreed between them, which are based broadly on standardised industry general terms and conditions, but with variations to reflect custom and practice between the two named counterparties concerned. OTC deals are bipartite so each party is exposed to the risk that the other party will default on performance or payment.

A swap allows a party to change its oil price exposure or risk from "floating" prices to "fixed" prices, or vice versa.

For example, if Company X had wanted to lock in the price of its crude oil or refined product acquisitions in year Y at, say, $105/bbl it might have bought an OTC swap from a bank at a fixed price of $105/bbl some time in year Y-1 or year Y-2. For simplicity let's say that Company X was able to buy the year Y swap at $105/bbl on 30th June year Y-1.

On 1st January year Y+1 the average actual oil price for year Y could be calculated because then all the prices from year Y would have been published. The average of these will be the floating price used to cash-settle the fixed price swap. The swap buyer effectively bought at the fixed price on 30th June year Y-1 and sold back at the floating price over year Y.

If that floating average price in year Y had turned out to be $120/bbl, the swap would cash-settle by selling to the bank at $120/bbl. If the hedge had been structured ideally, i.e. there is no basis risk, Company X will also have purchased its physical crude oil or refined products that it is hedging at the same floating price used to cash settle the hedge swap, i.e. $120/bbl. The net outcome for Company X, taking its physical oil purchases and its swap trade into account, would have been $105/bbl, as shown in Figure 26.

Figure 26 Hedging with an OTC Swap –Outcome 1

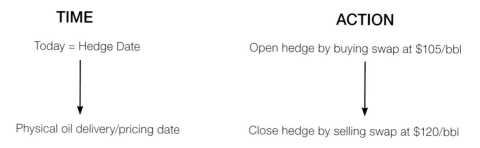

| TIME | ACTION |
|------|--------|
| Today = Hedge Date | Open hedge by buying swap at $105/bbl |
| Physical oil delivery/pricing date | Close hedge by selling swap at $120/bbl |
| | Buy Physical cargo at $120/bbl |

**Result:**

**Net overall purchase price = Physical purchase price ($120/bbl) . The effective purchase price is reduced by a hedge profit of $15/bbl, i.e. $120-15/bbl= $105/bbl.**

If on 1st January year Y+1 the average actual oil price in year Y had turned out to be $70/ bbl, the swap would cash-settle by selling to the market maker at $70/bbl. Again if the hedge had been structured ideally, Company X would at the same time have purchased its physical crude oil or refined products also at $70/bbl. The net outcome for Company X, taking its physical oil purchases and its swap trade into account, would once again be $105/bbl as shown in Figure 27.

In any and all cases the hedge has given Company X the perfect outcome: on 30th June year Y-1 Company X wanted to lock in its purchase price in year Y at $105/bbl. By buying the $105/bbl swap that is what it has actually achieved.

Figure 27 Hedging with an OTC Swap –Outcome 2

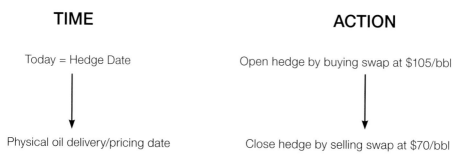

| TIME | ACTION |
|------|--------|
| Today = Hedge Date | Open hedge by buying swap at $105/bbl |
| Physical oil delivery/pricing date | Close hedge by selling swap at $70/bbl |
| | Buy Physical cargo at $70/bbl |

**Result:**

**Net overall purchase price = Physical purchase price ($70/bbl). The effective purchase price is increased by a hedge loss of $35/bbl, i.e. $70+35/bbl= $105/bbl.**

In the first case, when the actual price turned out to be $120/bbl, Company X ought to have been happy because it would be making money on the swap side of the equation, but it would by definition have been losing on the physical side. In the second case, when the actual price turned out to be $70/bbl, Company X ought also to have been happy because although it would be losing money on the swap side of the equation, it would be gaining on the physical side. However, Company X may have been less happy in the latter case, even though the financial outcome in both cases is exactly the same. This is because in the second case it would be paying money to a bank and would probably be wishing that it had not entered into the swap.

Companies that think this way, or have shareholders who think this way, may wish to consider if they would be "better off" not hedging. Unless the board of directors and the shareholders are comfortable with the purpose of hedging, and unless the risk profile, risk appetite and market price view have been considered and explained upfront, the risk management department may well be making a rod for its own back.

The swap pay-out curve for a hedger who buys a swap is shown in Figure 28. If the market price turns out to be above $105/bbl the hedger is making money on the swap. If the market price turns out to be below $105/bbl the hedger is losing money on the swap. In either case when the swap and physical position are taken together, the overall financial result for the hedger is that it is purchasing its oil at a net $105/bbl.

**Figure 28 Swap Buyer's Pay-out**

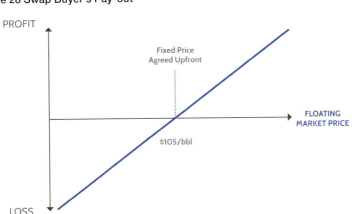

The swap seller's pay-out, typically an oil producer, would be the exact mirror image of the buyer's pay-out. (See Figure 29)

By entering into a swap any hedger that is a net buyer of oil, i.e. a consumer, has given up the chance of achieving a better net overall result than $105/bbl if the price falls after the swap has been purchased. Hedging the price of oil using a swap instrument locks in a price with certainty. This means that the hedger gives away all the upside if prices subsequently move in its favour[54].

**Figure 29 Swap Sellers Pay Out**

## Options

Like swaps, OTC options are typically, but not exclusively, used for strategic hedging purposes. Options allow the hedger to limit the downside from an adverse oil price move, but still enjoy the upside from a favourable price move. Options are also traded on regulated exchanges.

The **buyer** of an option is given the right, but not the obligation, to buy or sell oil at a specified price on a specified date. The **seller** of an option takes on the obligation, but not the right, if the option is exercised, to buy or sell oil at a specified price on a specified date. Buying an option is a risk-reducing activity. Selling, or writing, an option is a risk-increasing activity.

Two basic types of option exist:

- **Put**: The holder (buyer) of the option has the right, but not the obligation, to **sell** the

---

[54] The same is true for OTC forward and regulated future contracts.

oil at a given (strike) price, by (or on) a certain date (expiry);

- **Call**: The holder (buyer) of the option has the right, but not the obligation, to **buy** the oil at a given (strike) price, by (or on) a certain date (expiry).

For example, the buyer of a put option with a $105/bbl strike price acquires the right, but not the obligation, to **sell** the oil at $105/bbl. If the market price at the relevant period turns out to be $70/bbl then the holder of the option has the right to sell at $105/bbl by exercising the option. If the price turns out to be $120/bbl, the holder of the option does not have the obligation to sell under the option at $105/bbl. So it can sell its physical oil in the market at $120/bbl instead.

To acquire this flexibility to choose the better of the option strike price or the market price, the buyer of the option pays the seller of the option a premium upfront. Whether or not the buyer of the option decides to exercise it, the upfront premium for the option is a sunk cost that the buyer of the option does not get back.

The pay-out curve for the buyer of a put option is shown in Figure 30. The lower the actual market price turns out to be, the more profit the buyer of the option makes on the option. If the price moves higher, the option is allowed to lapse and the option buyer sells its oil in the physical market at the higher price. Put options are most often bought by oil producers.

**Figure 30 Put Option Pay-Out Curve**

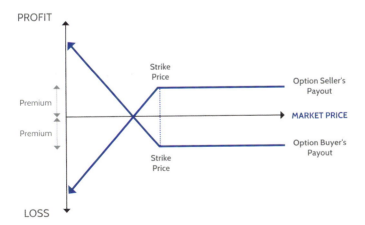

The lower the actual market price turns out to be, the larger the loss the seller of the option makes on the option, because it is obliged to buy at a higher than market price. If the price moves higher, the option is allowed to lapse and the option seller makes no profit, other than the premium it has been paid upfront by the buyer of the option.

Figure 31 shows similar pay-out curves for the buyer and seller of a call option.

**Figure 31 Call Option Pay-Out Curve**

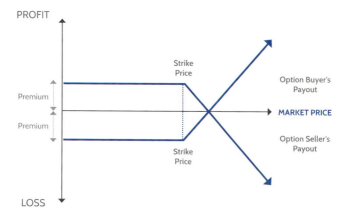

## Premium

The amount of premium that the option buyer must pay to the option seller is determined broadly by:

- the strike price of the option;

- the time value of the option;

- actual volatility of the oil price;

- interest rates; and,

- implied volatility.

**The strike price** of the option compared with the current underlying oil price for the period determines if the option has intrinsic value at the time the deal to buy the option is struck. If the current underlying market price is $95/bbl, a call option with a $100/bbl strike price will be worth considerably less than a $100/bbl strike price put option. This is because the put option has $5/bbl of intrinsic value. In other words the put option allows the holder of that option the right to sell at $5/bbl more than the current market price of $95/bbl. In other words the put option is **"in-the-money"**. The call option is **"out-of-the-money"**, i.e. it endows the option buyer with the right to buy at $5/bbl above the current market price. An out-of-the-money option will not be exercised. So while the market price is at $95/bbl the call option buyer would not exercise its right to buy oil at $100/bbl. This call option still has value even though it is currently out-of-the-money, because it may become in-the-money at some point before the option expires.

**The time value** is how long the buyer of the option is given to decide whether or not to exercise the option before it expires. The longer the time until expiry, the more expensive the option because the greater the likelihood that the market price will move to place the option in-the-money before it expires.

**Actual volatility** of the oil price has an impact on the option premium. The more volatile the oil price is, the more likely it is that the option will be exercised, so the higher the volatility, the higher the option premium. Oil prices can be very volatile (See Figure 32), much more volatile than interest rates or currencies and this can make oil options expensive to purchase outright. Other energy commodities are equally or even more volatile than crude oil.

**Figure 32 Historic Oil Price Volatility**

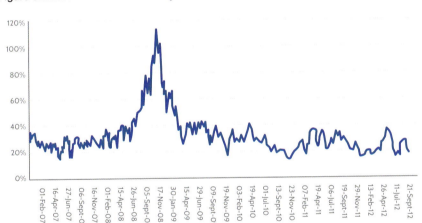

The higher **the interest rate,** the more an option grantor will want to compensate it for the risk of granting an option. This is because writing or selling an option is an investment on which the seller hopes to make a return. All investments are measured against the low-risk alternative, i.e. instead of writing an option, the option seller could put its money in the bank and earn interest. So the higher the interest rate, the higher the option premium.

**Implied volatility** can be defined as the level of volatility required to generate an option premium given a known market price for the underlying oil, a known interest rate, a known expiration date and a known strike price. It is short-hand for all other factors that have an influence on the premium. It may be determined by factors such as market liquidity or the credit risk associated with the particular counterparty concerned or the imminence of an OPEC meeting that may have an impact on oil supply. Anything that has an impact on market sentiment is picked up by implied volatility.

## The Option Style

Options can be categorised as European, American or Asian-style in accordance with the methodology by which they can be exercised. A European style option can only be exercised on its expiry date. An American style option can be exercised at any point up to its expiry date and is therefore a more flexible option than its European cousin. An Asian style option is automatically exercised or automatically allowed to expire depending on whether the average market price over the option settlement period is above or below the strike price of the option.

Most, but by no means all, options in the crude oil market tend to be Asian style options that are either exercised or not exercised based on whether the strike price is above or below a monthly average published price or sometimes a quarterly average published price.

So, for example, let's say a producer buys a $100/bbl Asian-style put option with a monthly settlement in year Y. If the monthly average price in January Y turns out to be $95/bbl, the producer will automatically receive $5/bbl from the option seller. If the price in February Y turns out to be $105/bbl, the option for that month will expire worthless and no money will change hands between the put option buyer and the put option seller. The March Y through December Y put options will be settled in the same way at the end of each month depending on whether the monthly average published price turns out to be above or below $100/bbl.

This Asian expiry method more closely tracks the way physical contracts are priced in the crude oil market than do the European and American style options. Long term physical crude oil contracts usually contain monthly average price formulae, so hedging with an Asian style option reduces basis risk. However, if the hedger does not enter into term physical contracts, but instead deals in the spot market using 3-5 day average price formulae related to the B/L date of each cargo then it will have basis risk against a monthly average hedge settlement price. In these circumstances the hedger will again have to enter into a second layer of hedging of the time differential, T, using the CFD market as mentioned above in connection with hedging with swaps.

In fact most options are never exercised, but are sold before the expiry date. This is because if the option has intrinsic value, i.e. it is in-the-money, that intrinsic value can be valorised by selling the option. If, as an alternative, the option is simply exercised the remaining time value in the option will be lost.

For example, say a hedger has purchased a put option with a $100/bbl strike price and the market price falls to $90/bbl. The option now has an intrinsic value of $10/bbl. The option holder can exercise the option to sell at $100/bbl, i.e. $10/bbl above the current market price. This will give it a profit of $10/bbl. If instead the option holder sells the option rather than exercising it, it will receive at least the $10/bbl of intrinsic value, plus additional premium to reflect the remaining time value in the option.

When companies start hedging for the first time it is often pointed out to them that hedging is akin to taking out an insurance policy. In the case of hedging with forwards, futures and swaps the "insurance premium" amounts to the entire potentially favourable price movement that the market might make after the hedge has been put in place. Sacrificing all the price upside to protect against the price downside may not suit the risk appetite and market price view of every company.

In the case of purchasing put or call options to protect against falling or rising prices respectively, the analogy with insurance is much closer. The hedger that implements an options strategy is protected against all its exposure to adverse price moves, but retains all the benefits of favourable price moves after the option has been purchased. The only money the hedger puts at risk is the upfront premium, which is retained by the option seller whether the option is exercised by the option buyer or not.

The insurance analogy is very attractive to the new hedger up until the point when the size of the option premium is discussed. As mentioned above OTC options are favoured by strategic hedgers so almost by definition the option programme will be undertaken in significant size and for a long duration, i.e. with a high time value. For example, at the time of writing this book the cost of a put option that is $5/bbl out-of-the-money with an expiry date one year forward has a premium close to $10/bbl. In other words it costs $10/bbl upfront for the right to sell at $5/bbl below the current market price for oil for delivery in one year's time. Once the scale of a strategic hedging programme involving many millions of barrels is considered, the cost of insurance starts to look a bit daunting. The point at which simply buying put options becomes too expensive is an entirely subjective one.

But as with home insurance or car insurance if an "all risks" policy is too expensive a "third party, fire and theft" policy may be all that is needed. This is why oil hedgers often get involved in "exotic" options packages. Sometimes this has disastrous results if the possible consequences of the strategy are not explained clearly upfront to board members and shareholders.

## Option Strategy

Talking to options traders can be very confusing because of the bewildering terminology they use. In fact options are very simple. "Exotic" option packages offered on the market - straddles, strangles, ZCCs, seagulls, condors, knock-outs, targets etc. - are no more than a series of put option and call option building blocks put together into different and increasingly complex structures. But if the hedger understands how puts and calls work and it maintains a clear picture of its own risk profile, risk appetite and market price view while it selects the option package best suited to its needs, the rest is easy.

The premium charged for buying puts and calls outright can be considered by some to be prohibitive, particularly in volatile markets where price risk and the cost of options escalates for long-term strategic hedges. So hedgers will often to seek to finance the purchase of the options they need for hedging purposes by selling a different kind of option. Selling one option can raise the money to underwrite the purchase of another option.

It should be recalled that we said at the start of this section on options that buying an option is a risk-reducing activity, whereas selling, or writing, an option is a risk-increasing activity. Hedgers seeking to defray the cost of buying the option they need would be well advised never to lose sight of this fact.

Nevertheless selling options can be one way of financing risk cover, so long as the hedger has a clear idea of its risk profile, risk appetite and price view and understands the implications of an option being exercised against it.

Let's say an oil buyer wants to protect itself against the oil prices rising above $100/bbl in one years' time. Let's say the forward oil curve today shows the market in backwardation with the price of oil for delivery in one months' time being, say, $110/bbl and the price of oil for delivery in one years' time being, say, $95/bbl.

The hedger could buy the swap for oil for delivery in one years' time at $95/bbl and this would meet its objective of ensuring that it does not have to pay more than $100/bbl for oil for delivery in one years' time. However the hedger's market price view might be that there is a reasonable chance that the price of oil may fall to $90/bbl, in which case it does not want to be locked in to buying at a price of $95/bbl with a swap contract. It wants to have the opportunity to buy at $90/bbl even though the market does not currently offer that price.

The hedger could buy $100/bbl call option on oil for delivery in one years' time. Let's say the premium it would have to pay for a call option that is currently $5/bbl out-of-the-money[55] (o-t-m) is $10/bbl. If the hedger wants to buy the $100/bbl call option it will have to pay $10/bbl upfront. That money is gone for good; regardless of what subsequently happens in the market that premium is now a sunk cost.

If the market price subsequently rises to, say, $130/bbl, the hedger will exercise its right to buy at $100/bbl and will have saved itself $30/bbl. However, when the cost of the upfront premium of $10/bbl is taken into account its net purchase price is $110/bbl, i.e. the strike price of $100/bbl plus the premium of $10/bbl. The hedge has not been entirely successful because the hedger wanted to protect itself against having to buy at more than $100/bbl, but its net purchase price has turned out to be $110/bbl.

If on the other hand the market price subsequently falls to $90/bbl, the hedger does not exercise its call option to buy at $100/bbl and simply buys in the market instead at $90/bbl. However, when the cost of the upfront premium of $10/bbl is taken into account its net purchase price is $100/bbl, i.e. the market price of $90/bbl plus the premium of $10/bbl.

## The Zero Cost Collar

The cost of buying options is the reason why zero cost collars (ZCCs) are often found acceptable by oil firms. As the name implies the hedger does not need to find any upfront premium. A ZCC strategy requires the hedger to ask itself how much is the option I want to buy going to cost me? What option can I sell that generates enough money to pay for the option I really want to buy? If the option that I sell is exercised against me is the action I will have to take in keeping with my current risk profile, my current risk appetite and my current market price view?

The other question the hedger should be asking is, can I buy the option I want from another derivatives provider at a cheaper price? Shopping around is invariably a good idea, but that is not as easy as it sounds. It may be that the hedger is required to hedge by its bank in order to underwrite a loan. The bank may put pressure on the hedger to

---

[55] In other words the call options strike price is $100/bbl, which is $5/bbl above the current market price for oil for delivery in one years' time at $95/bbl.

buy its derivatives from the bank's own affiliate. Even if it does not, there are barriers to entry in the derivatives market, one of which is getting credit clearance to deal with banks and another is the need to sign weighty International Swaps and Derivatives Association (ISDA) master contracts with a range of derivative providers.

Figure 33 demonstrates the structure of a ZCC that would be appropriate to a net buyer of physical oil. The oil buyer buys the $100/bbl call option and finances this by selling a $90/bbl put option. Hence it pays no upfront premium.

**Figure 33 Oil Buyer's ZCC Pay-Out Curve**

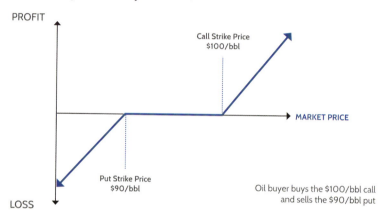

The ZCC now provides the oil buyer the protection it wanted of never having to buy at more than $100/bbl. The oil buyer also wanted the opportunity to buy at $90/bbl if the price fell. Now it has the *obligation* to buy at $90/bbl if the price falls to $90/bbl, or *lower*. If at the outset when the oil buyer was considering its hedge strategy the opportunity to buy at $90/bbl had been on offer in the market, the oil buyer may have chosen to buy a $90/bbl swap and not undertaken any options at all. So *in theory* the hedger should be happy with its ZCC outcome even if the price falls to $70/bbl. Inevitably, as discussed above, if the price falls to $70/bbl there will be some grumbling that the company would have been better off not hedging at all.

For completeness let's look at the same issue from the perspective of an oil producer. The producer may be concerned to protect itself against the risk of the price of oil for delivery in one years' time falling below $90/bbl. Let's say we are facing the same forward oil curve today as before with the market in backwardation. The price of oil for delivery in one months' time is $110/bbl and the price of oil for delivery in one years' time is $95/bbl.

The producer can fix its one year forward sales price at $95/bbl by selling the $95/

bbl swap. But it may wish to retain the opportunity to sell at $100/bbl if the price rises. So it may consider buying a $90/bbl put option, which will give it the right, but not the obligation to sell at $90/bbl if the price falls to, say, $80/bbl. Let's say the premium for a $90/bbl put option, i.e. an option that is $5/bbl out-of-the-money one year forward, is $7.50/bbl.

### An Aside on Option Valuation

*An obvious question that may be asked is if the $5/bbl o-t-m call option 1 year forward costs $10/bbl, why does the $5/bbl o-t-m put option 1 year forward cost $7.50/bbl? It is not the task of this book to explain options valuation: that would require a whole different book discussing the value of delta[56] and gamma[57] and skews[58] and it would not be written by this author.*

*For our purposes, suffice to say that risk in an option is neither symmetrical nor linear. In both the call and put options referred to in our example the relationship between the strike price and the market price is the same, i.e. $5/bbl o-t-m. The time value is also identical as is the interest rate. The historic volatility appears to be the same too. (Although in reality it is not because of a skew factor). But the implied volatility is very different.*

*Let's take a practical example to explain why this may be so. Let's assume that the same market-making derivative provider is approached by an oil buyer and an oil seller simultaneously, each wanting to buy a $5/bbl o-t-m option one year forward. The current price is $95/bbl. The buyer wants to buy the $100/bbl call option and the oil seller wants to buy the $90/bbl put option. Let's say the derivative provider is already very short of the market, i.e. the market- maker has already sold a lot of oil and will have to buy in the market at some point in order to balance its position.*

*If the market-maker sells the call option to the oil buyer it may be exercised against the market-maker and it will be forced to sell oil (i.e. the buyer of the call option has the right but not the obligation to buy, the seller of the option has the obligation but not the right to sell). This will make the market-maker even shorter still, which may be a position that it does not relish. So it will charge quite a high price for a call option.*

*If the market-maker sells the put option to the oil seller it may be exercised against the market-maker and it will be forced to buy oil (i.e. the buyer of the put option has the right but not*

---

[56] Delta is the rate of change in the price premium paid for an option as the market price of oil changes.

[57] Gamma is the rate of change in the delta of an option as the market price of oil changes.

[58] O-t-m options are traded more actively than At-the-Money (a-t-m) options and the greater turnover leads to greater volatility. Furthermore the market may have a view that if prices go beyond a certain level then the fundamentals have changed and so therefore has volatility. So different strike prices have a different range of volatilities applied to them. This is known as skew.

*the obligation to sell, the seller of the option has the obligation but not the right to buy). This will make the market- maker less short, which suits its book. So it will charge a lower price for a put option.*

If the oil seller buys the $90/bbl put option and the price falls to, say, $80/bbl it will exercise its option and sell at $90/bbl. But once the cost of the upfront premium to the hedger of $7.50/bbl is taken into account the oil seller's overall net realisation is $82.50/bbl. The hedge has not been entirely successful because the hedger wanted to protect itself against having to sell at less than $90/bbl, but its net sales price has turned out to be $82.50/bbl.

The seller may decide to finance the purchase of the $90/bbl put option by selling an o-t-m call option at, say, $100/bbl[59].

The overall effect is as demonstrated in Figure 34 below.

**Figure 34 Oil Seller's ZCC Pay-Out Curve**

The ZCC now provides the oil seller the protection it wanted of never having to sell at less than $90/bbl. The oil seller also wanted the opportunity to sell at $100/bbl if the price rose. Now it has the obligation to sell at $100/bbl if the price rises above $100/bbl, *or higher*. If at the outset when the oil seller was considering its hedge strategy the

---

[59] Those who are following the development of this example closely will want to know why the oil seller, who wants to find $7.50/bbl in order to buy a $90/bbl put option, has to sell a $100/bbl call option, which we said above cost the oil buyer entering into a similar hedge $10/bbl. Surely the oil seller will be able to sell call options at a greater strike price than $100/bbl for a lesser premium of $7.50/bbl? However it should be borne in mind that the market maker was prepared to sell the $100/bbl call option at $10/bbl, but that does not mean to say that the same market maker will be prepared to buy the $100/bbl call option at $10/bbl. There will be a bid-offer spread applied by the market maker, which will vary with its overall book and its profit margin. The prices it quotes to individual hedgers will also reflect its assessment of that company's credit risk and its ability to shop around a range of its competitors for a better price.

opportunity to sell at $100/bbl had been on offer in the market, the oil seller may have chosen to sell a $100/bbl swap and not undertaken any options at all. So in theory the hedger should be happy with its ZCC outcome even if the price rises to $120/bbl. Inevitably, as discussed above, if it does there may well be questions asked by board members or shareholders.

ZCCs are tidy options because they are easy to model and explain and since no upfront premium is required there is no immediate impact on the cash-flow of the company. They are also lazy options because a bit more effort on the part of the hedger to analyse its risk profile, risk appetite and market price view can deliver a better options package more tailored to its own needs.

## Tailored Option Packages

When an oil firm is devising the corporate strategy for its usual planning horizon it may, as discussed above, come up with a scenario plan that models its high, low and base case price assumptions. It may go as far as assigning probabilities to the different price outcomes and have contingency plans in place to cope with the least desirable of these.

Let's say the scenario plan envisages a base case assumption of $100/bbl with a low price case of $60/bbl and a high price case of $150/bbl.

Typically the oil company will assign low probabilities to the high and low price cases and a higher probability to the base case scenario. Usually the base case assumption lies at or near current oil price levels. Any hedging action the company considers will probably be in a range around the base case assumption as being outcomes that have a higher probability of coming to pass than have the high or low cases.

In all likelihood it will consider hedging a fixed proportion of its forward position, say 25-50%: it would be quite uncommon for an oil producer to hedge 100% of its forward production profile or for a refiner to hedge 100% of its likely feedstock requirements.

If an oil producer were given the opportunity to sell at its high case price assumption when it is constructing its scenario plan it may well feel motivated to sell a large proportion of its production at this high price, which is unattainable at that time. The high case price assumption of $150/bbl has to be unattainable or else it would not be the company's high price case.

In theory then when the producer is constructing its plan and settling on a base case assumption of $100/bbl it may think that it would be happy to sell all its oil at $150/bbl if the opportunity arose. Since it cannot do that it could instead sell $150/bbl call options and collect upfront premium. Selling a call option gives the producer the obligation to sell at $150/bbl if the market ever reaches that level by granting a third party the right but not the obligation to buy at $150/bbl. If the market price goes to $170/bbl the producer will be obliged to sell at $150/bbl. But this was its high case price assumption at the outset, so by definition the oil producer should be in clover if it is forced to live up to this obligation. At $150/bbl it can carry out all the projects it had in mind in its corporate plan.

If the price falls to $90/bbl the oil producer's base case price assumption is not protected, but it at least has some option premium from the sale of $150/bbl call options to cushion the blow. But that is not the end of the story.

When it sells the $150/bbl call options the oil producer in all likelihood would use that premium income to protect its base case price assumption. It may buy $90/bbl put options and defray the cost of this acquisition using its $150/bbl option premium revenue. The premium income from the sale of $150/bbl calls will be quite low because the options are very o-t-m, i.e. the option strike price is well above the current market price.

However it may be that because $150/bbl is so high and that price outcome has such a low probability attached to it, in the company's price view, that it may be prepared to sell two $150/bbl calls to defray the cost of purchasing one $90/bbl put. Even then the 2-for-1 option package may not be zero net cost. In other words the oil producer may still have to pay some money for the $90/bbl put option it wants. But the company may judge that some cost is acceptable to protect itself against a price fall below $90/bbl, which has a higher probability in its view than does the $150/bbl high price case. Figure 35 models the pay-out curve for a 2-for-1 leveraged collar with a small net premium.

**Figure 35 Oil Seller's Leveraged Collar with Small Premium Pay-Out Curve**

Similarly in its scenario plan the oil producer has assigned a low probability to its low oil price case of $60/bbl. If it did not assign a low probability to this case, then it would not be the company's low price case.

So if the oil producer wants to have protection against the price falling below $90/bbl, which it considers quite likely, should the oil producer be prepared to defray the cost of buying $90/bbl puts by selling $60/bbl puts? Selling a $60/bbl put would give it the obligation to buy at $60/bbl if the price fell below that level. In fact the oil producer would not be a net buyer of oil if the put option were exercised against it: it would simply cancel the hedge protection it put in place at $90/bbl when it bought the $90/bbl put option and sold the $105/bbl call option. It is now simply buying back that $90/bbl protection at $60/bbl. Figure 36 models the situation of an oil producer that buys a ZCC at $90-105/bbl then re-opens downside risk below $60/bbl. This strategy is referred to as a "seagull" for obvious reasons.

**Figure 36 Oil Seller's Seagull Pay-Out Curve**

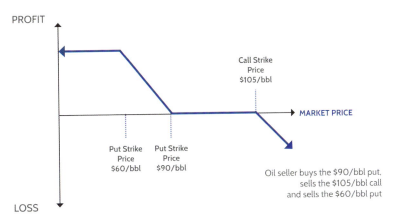

By eliminating "unnecessary" downside protection against a price fall below $60/bbl, which the producer deems an unlikely outcome, the producer has increased the ceiling in its ZCC from $100/bbl to $105/bbl. In other words it has gained the exposure to an extra $5/bbl of upside, which it considers quite likely to occur, by foregoing protection against an event that it considers unlikely, i.e. that prices will fall below $60/bbl.

At what point does the options package become so complex that the hedger has crossed the line and become a speculator? Only the hedger concerned can know because any action may be hedging or speculation depending on the company's risk profile, risk appetite and market price view at the time it undertook each step.

There is only one "bad" outcome from investing in an options package: a surprise. The outcome of a hedge, options-based or otherwise, should never be a surprise to the board or to the shareholders, otherwise the implementer of the hedge is leaving the door open for a witch hunt. A hedge should never be undertaken unless the company understands what it is doing, why it is doing it and what the outcome could be in a worst case scenario.

Derivative providers are required by law to know their customers and to receive direct confirmation from the board that the individuals implementing hedges are properly authorised to take action and that the particular action they take is sanctioned by the board. Nevertheless "when good hedges go bad" it is not unusual for the board to instigate a change of personnel and claim that the previous incumbents were acting without authority.

Any oil company directors that have struggled with this Chapter might be well-advised to re-read it and get fully comfortable with its contents before sanctioning a strategic hedging programme in their own companies. It is better to ask the basic questions before the company takes action than to have them questioned by the financial press or in court. The concepts described above are not difficult; they just require a little bit of effort to understand. Without that understanding companies may wish to avoid hedging altogether.

# A Final Word

In this book we have tried to explain the peculiar complexity to a newcomer of the international crude oil market. We have conducted our analysis in terms of the three, now familiar, components of the oil price: the absolute price (A); the time differential (T); and, the grade differential (G). We recognise that traders do not use this Consilience terminology in transacting their deals and it is typically only when they decide to fix values upfront, or when they unbundle the price to manage the elements separately by hedging, that the three components emerge clearly.

We have looked in detail at the benchmark grades of oil used to establish the value of the absolute price, A, concentrating on Brent, WTI and Dubai and highlighted the shortcomings of these three marker grades. We have noted with concern the vast and growing quantity of oil that is priced by reference to the declining physical production of the Brent benchmark, not only in the physical market but in the forward, futures and OTC derivative markets. We have likened Brent to an inverted pyramid with, at the very least, 200 billion bbls per year balancing shakily on an unstable point of a reducing production and sparse deal evidence. Changes are being discussed and stakeholders, may well wish to make their voices heard in the debate over how to preserve this key benchmark grade. Despite its shortcomings Brent is still the best benchmark we have.

We have examined the pros and cons of using forward and futures contracts in the operational hedging of the absolute price, A, in the short term. From a long term strategic viewpoint we have explored the use of the OTC swaps and options markets to manage the value of A and described how to add this ordinance to the corporate arsenal without blowing off the company's metaphorical foot.

We have analysed the value of the time differential, T, and noted how it can become blurred in price negotiations to give a totally mistaken apparent value of the grade differential, G. This can mislead small oil producers and refiners into believing that they are achieving better relative values in their crude oil sale and purchase contracts than may in fact be the case. We have pointed out the danger of having a price in a sales contract that diverges from the price at which costs are recovered, profit share is calculated and tax is levied upstream.

We have studied the value of the time differential, T, and saw how this value can be managed in the dated-to-paper CFD markets. This market is an essential tool for fine-tuning operational hedges and for perfecting strategic hedges as they mature over time.

We have also examined the crude oil grade differential and the impact that quality, refined product yield, location and terminal logistics can have on the value of oil. We have recommended addressing these factors early in field development plans to ensure that the economic model reflects accurately the different price and operating cost implications of alternative field designs.

When I joined the oil industry in 1978 life was a lot simpler. Oil price volatility was considerably less and contract prices were transparent and straightforward. The commoditization of oil progressed by a series of quantum leaps with the rise of the spot market, the emergence of forward and futures contracts and the development of a range of OTC swap and option tools. The backdrop to each change was an increase in price volatility that had to be managed.

Regulation being considered in the wake of the banking crisis aimed at cutting down on inappropriate risk-taking by banks, weeding out market manipulation and increasing transparency in the OTC markets requires careful consideration in its potential application to the oil market.

Although oil has been commoditized over time it is not a homogenous commodity. Managing price risk involves huge basis risk because the financial instruments available are not sufficiently diverse or liquid to be up to the task. If the banks, starting with US banks, are prevented from proprietary trading they will be unable to take on this basis risk, so it will have to stay with the oil industry where it is not welcomed by independent producers, refiners and end-users, who might be unable to cope with it.

This has implications for the role of the independent company who may find it increasingly difficult to compete with the more financially established major oil companies, NOCs and trading houses.

A shoe-horning of highly tailored risk instruments onto regulated exchanges is a cure worse than the disease. The cash margin payment demands of clearinghouses are beyond the reach of many of the hedgers for whom price risk management is an essential prerequisite of guaranteed development financing.

The most flexible, liquid and ubiquitous suite of price-setting and price risk management instruments available in the market remain based on Brent. But the structure is at risk and a wrong move in preserving an appropriate physical base can bring the whole structure of the price formation process crashing down.

While some may welcome a return to a simpler age of straightforward fixed and flat prices and long term contracts, oil price volatility is here to stay. So regulating away the instruments designed to manage that volatility would be a retrograde step. We can't un-invent the wheel.

# Appendix 1: Breaking News on Brent

In Chapter Three of this book we estimated conservatively that in excess of 200 billion bbls/year of international oil contracts – physical, forwards, futures, swaps and options- are priced by reference to Brent and said that the physical production of Brent is on a declining trend. We described the moves that have been made in the past to support the physical base underpinning the Brent suite of contracts: the inclusion of trades in physical UK and Norwegian Forties, Oseberg and Ekofisk cargoes into the assessment process for the influential Dated Brent published price that is reputed to be used in two thirds of the world's physical contracts; and, the inclusion of a provision to supply Forties, Oseberg or Ekofisk as a substitute for Brent in the forward 25-Day BFOE contract.

We said that further changes are likely to be necessary to extend the life of this flawed benchmark grade because production of all four deliverable grades continues to decline and there is no successor benchmark emerging as heir apparent. We noted that one of the market price reporting agencies, Platts, indicated in October 2012 that it was considering including additional grades of crude oil from a wider geographic area in its assessments of Dated Brent and of the 25-Day BFOE contract.

On 8th February 2013 Shell pre-empted any such move by Platts and announced that with effect from 11th February it would change the terms on which it would trade 25-Day BFOE cargoes for delivery in May 2013 and beyond. The mechanism chosen by Shell to implement this change was the issuance of an amendment to its "SUKO 90" terms[1]. These are the General Terms and Conditions of Trade (GTCs) that have been adopted by all players trading in the 25-Day BFOE market. Shell cannot impose these amendments on any other company that chooses to use the 25-Day BFOE market and has made no attempt to do so. It has simply said that these are the terms on which it will trade henceforth. BP followed suit rapidly and adopted the Shell amendments for its own trades.

The problem that the Shell amendments are attempting to fix is the fact that sellers of 25-Day BFOE cargoes tend to always deliver Forties cargoes into the contract because, following the introduction of the low quality Buzzard field into the Forties Blend in 2007, Forties Blend tends to be the cheapest of the four grades - Brent, Forties, Oseberg and Ekofisk - that qualify for delivery into the forward contract. However the availability of Forties cargoes in the market is unpredictable. This is caused by variable and declining production levels. The steady sale of VLCC-sized[2] Forties cargoes to South Korea each month and a constant appetite for pipeline transfers of Forties Blend into the Grangemouth refinery adjacent to the Forties loading terminal at Hound Point reduces the number of cargoes available to supply into 25-Day BFOE chains. This had made the price of Forties Blend increasingly volatile introducing an unmanageable source of basis risk for those trading and hedging using the 25-Day BFOE contract.

The Shell solution has been to introduce a series of price escalators to apply to deliveries of Brent, Oseberg and Ekofisk to increase the chances that sellers of 25-Day BFOE

---

[1] Agreement for the Sale of Brent Blend Crude Oil on 15 day terms Part 2 General Conditions Shell U.K. Limited July 1990

[2] VLCC stands for Very Large Crude Carrier. This is a tanker that can load between 2 to 2.2 barrels.

cargoes will actually deliver one of the alternative grades instead Forties. The calculation of the escalator is quite complex but it allows sellers of 25-Day BFOE contracts to establish the escalation factor that would apply to the contract price of each grade before the grade declaration has to be made to the buyer 25 days before the cargo loads.

The amendments introduced by Shell are reproduced below and they require some explanation.

*"Where a Seller delivers a Cargo of Brent, Ekofisk or Oseberg Crude Oil pursuant to a contract governed by these General Terms and Conditions for delivery in May 2013 or thereafter under these terms, the price for such cargo shall be adjusted by a Quality Premium which shall always be positive or zero.*

- *If a Cargo of **Brent** is declared then the Quality Premium shall be equal to the larger of zero and:*

  » *Twenty five percent (25%) of one sixth (1/6th) of the sum of three (3) times the Brent Differential less the Forties Differential for M-2 plus; two (2) times the Brent Differential less the Forties Differential for M-3 plus; the Brent Differential less the Forties Differential for M-4.*

- *If a Cargo of **Ekofisk** is declared then the Quality Premium shall be equal to the larger of zero and:*

  » *Fifty percent (50%) of one sixth (1/6th) of the sum of three (3) times the Ekofisk Differential less the Forties Differential for M-2 plus; two (2) times the Ekofisk Differential less the Forties Differential for M-3 plus; the Ekofisk Differential less the Forties Differential for M-4.*

- *If a Cargo of **Oseberg** is declared then the Quality Premium shall be equal to the larger of zero and:*

  » *Twenty five percent (25%) of one sixth (1/6th) of the sum of three (3) times the Oseberg Differential less the Forties Differential for M-2 plus; two (2) times the Oseberg Differential less the Forties Differential for M-3 plus; the Oseberg Differential less the Forties Differential*

*The Differential shall be the average of the low and high assessments for "Spread vs fwd Dated Brent" as quoted in the Platts Crude Oil Marketwire for all quote publication days in the applicable month for the applicable grade."*

## Which Differential or "Spread"?

The starting point of the Shell calculation is the published price assessment of the "spread versus forward Dated Brent". This data is published by Platts and is derived from the forward price curve, specifically the CFD prices described in Chapter Four. This data shows the anticipated value today of what the price differential between the various grades and Dated Brent will be 10-25 days forward of today's date. So the data used by Shell is today's assessment of what the grade differential will be, not what it actually is today.

The price differential today between the various grades and Dated Brent relates to cargoes loading in the next 10-25 days. Instead the "spread versus forward Dated Brent" used by Shell employs today's assessment of what that price differential will be in 10-25 days' time for cargoes loading 10-25 days after that.

# The Differential Between Differentials

As a further complication Shell takes this "spread versus forward Dated Brent" data **and deducts:**

- The Forties spread versus forward Dated Brent from the Brent/Ninian Blend[3] spread versus forward Dated Brent; and,

- The Forties spread versus forward Dated Brent from the Ekofisk Blend spread versus forward Dated Brent; and,

- The Forties spread versus forward Dated Brent from the Oseberg Blend spread versus forward Dated Brent.

This gives us today's assessment of how much more expensive than Forties the three alternative grades are anticipated to be in 10-25 days' time.

# The Averaging Period

In order that the seller can calculate what premium it will receive if it delivers Brent Blend or Ekofisk or Oseberg into the 25-Day BFOE contract the differentials described above are calculated as an average over a historic period. For the 25-Day BFOE delivery month M, the averaging is carried out on data published over M-4, M-3 and M-2.

So, for example, for 25-Day BFOE cargoes for delivery in March (M) the average of the relevant differential over November (M-4), December (M-3) and January (M-2) is calculated. So that by the time the seller begins to "wet chains" in early February (M-1), by the process described in Chapter Three, there will be full price discovery of the premia that will apply to each of the grades that might be supplied as an alternative to the cheapest one, i.e. usually Forties.

---

[3] Brent/Ninian Blend, or BNB, is the blend that loads at Sullom Voe in the Shetland Isles and is the name given by Platts to the original Brent. The price of BNB differs from the price of what is now called Dated Brent, which is the price of the cheapest of BNB, Forties, Oseberg or Ekofisk.

# The Weighting of the Average

The average calculation is not a simple average. The most recent data, i.e. the data for M-2, or January in our example, is given a weighting factor of 3. The data for M-3, or December in our example, is given a weighting factor of 2. The data for M-4, or November in our example, is given a weighting factor of 1. The three averages are summed and divided by 6 to give a weighted average.

# The Fudge Factor

Once these weighted average premia have been calculated, as described above, an arbitrary further scaling factor is applied to each of the three alternative grades of Brent Blend, Ekofisk Blend and Oseberg Blend. These are:

- 25% of the weighted average historic premium in the case of Brent Blend; and,

- 50% of the weighted average historic premium in the case of Ekofisk Blend; and,

- 25% of the weighted average historic premium in the case of Oseberg Blend.

In all cases if the premium calculation were to deliver a negative number, i.e. Brent, Ekofisk or Oseberg were to be at a discount to Forties, a lower limit of zero would apply to the premium. In these circumstances, i.e. if Forties were to be the most expensive grade, a seller would need no incentive to supply an alternative grade. For example, this might occur when the low quality Buzzard field is shut down for maintenance.

The extra premium given to Ekofisk, relative to that for Brent Blend and Oseberg Blend, i.e. 50% rather than 25%, does not reflect any obvious quality attributes. It appears to be an attempt to encourage the delivery of what is the second biggest stream of North Sea oil available into the ailing 25-Day BFOE contract.

# A Broken Benchmark

It is not immediately obvious why receiving a premium that amounts to a quarter of the market's assessment of what the future grade differential averaged over a historic three month period would encourage a 25-Day BFOE seller to deliver Brent or Oseberg rather than Forties into the 25-Day BFOE contract. Similarly why should receiving a premium that amounts to a half of the market's assessment of what the future grade differential averaged over a historic three month period encourage a 25-Day BFOE seller to deliver Ekofisk rather than Forties? If Forties is the cheapest grade available on the market, sellers will supply Forties.

But it is to be hoped that these premia will put a limit on the extreme oscillations in the price of Forties that have been witnessed when Buzzard has been shut in and which could become more violent as Forties Blend production declines and Buzzard forms a larger proportion of the blend.

## What Next?

Shell's move, closely followed by that of BP, is not the end of the story. These companies can only say how they will trade the 25-Day BFOE contract. They have no power to impose these terms on other market participants. It is in the interests of all participants in the market that the contract trades on standard terms. Otherwise buyers and sellers run the risk of a large price exposure if they buy on the new Shell terms and sell on the old terms that have no Brent, Ekofisk and Oseberg premia included.

There is likely to be some debate over whether Shell's somewhat convoluted premium calculation methodology is the right one and whether the 25%/50%/25% factors applied to Brent, Ekofisk and Oseberg are appropriate. But it is probable that other market participants will adopt the new terms rather than take on the unnecessary basis risk of not dealing on back-to-back terms when they are buying and selling 25-Day BFOE cargoes.

The problem is that until Platts agrees to include cargoes traded on the new Shell basis in its price assessment process the market runs the risk of two tier pricing. On 18th February Platts counter-proposed a very similar idea to the one proposed to Shell. Platts would like to see escalators based on an average of one month historic data, rather than three months, and applying only to Oseberg and Ekofisk, not to Brent. Platts' suggested escalation factor is 50% for both Ekofisk and Oseberg. This means we now have two alternative proposals and market participants are in a quandary about which way to trade until this unhelpful uncertainty is resolved.

How much easier life would be for this troubled, but highly influential, contract if all the interested parties could have sat down in one room and thrashed out the changes introduced by Shell in advance and could discuss the further changes that will be needed as the production of the four grades that currently form the "Brent" market decline further.

As we have said throughout this book that cannot happen without regulatory oversight of the physical base underlying this crucial crude oil benchmark, which sets not only the price of the majority of the world's physical oil contracts, but a significant proportion of the futures and derivative markets too. Any attempt by the industry to coordinate contractual changes of the type discussed in this appendix is considered by the major actors to open the door to accusations of price collusion.

# Table of Figures

## Tables

## Figures

# Glossary

| Term | Definition |
|---|---|
| 25-Day BFOE Contract | This is an OTC forward contract for the purchase and sale of 600,000 bbl cargoes of either Brent, Forties, Oseberg or Ekofisk in a future month. The seller is obliged to inform the buyer of which of the four grades of crude oil it will receive and on which three day loading date range the cargo will be delivered in the specified month by 4pm London time 25 clear days before the first day of the three day loading date range. |
| 2-1-2 | Used in price formulae to denote that the price index in question should be averaged over a 5 day period that is the two business days before the B/L date, the B/L date and the two business days after the B/L date. |
| A | The absolute price, represented by the height of the forward oil price curve. |
| Actual Volatility | The historic variability of the price of the underlying commodity on which an option has been written. |
| AFE | Authority for Expenditure. |
| American Petroleum Institute (API) | American Petroleum Institute founded in 1919 as a trade association for the oil industry. Publishes weekly information on US petroleum stock figures, refinery throughput, imports, exports and stock levels. This information is divided into five geographical areas known as PADDs. This stands for Petroleum Administration Defence Districts. The API also established the internationally recognised system for grading crude oils by specific gravity (API gravity). |
| API Gravity | A scale expressing the gravity of petroleum products and crude oil devised by the API and the National Bureau of Standards. The higher the API gravity the lighter the crude oil and vice versa. |
| APPI | The Asian Petroleum Price Index. A price reporting agency that has now ceased trading, but was influential in compiling prices that were employed by NOCs in the Far East in setting tax reference prices. |
| ARA | Amsterdam-Rotterdam-Antwerp area - a port and refining area in the Belgian-Dutch region. Crude oil may be delivered CIF ARA meaning that it is discharged in this area for refining. A product cargo bought on a FOB ARA basis means the oil can be supplied from any of these ports. |
| Arbitrage | Trading that profits from discrepancies in prices due, for example, to location, illiquidity, slow communication of new information, or any other reason. Traders will sell those prices, or price components, that are over-valued, driving the price down, and buy those price components that are undervalued, driving the price up, so that they rapidly come back into line. |

| | |
|---|---|
| Aromatics | Cyclical unsaturated hydrocarbon chains. |
| Assay | A laboratory analysis of quality. |
| ASCI | The Argus Sour Crude Index. This has been published by Argus Media since May 2009. It represents a volume-weighted average of deals in the US Gulf coast medium sour crude oil grades of Mars, Poseidon and Southern Green Canyon (SGC). The ASCI price is expressed as a differential to WTI. |
| Atmospheric Crude Oil Distillation Unit | See Distillation. |
| Atmospheric Pressure | The pressure of the air at sea level. |
| Backwardation | The forward oil price curve slopes down from left to right meaning that the price of oil for delivery tomorrow is more expensive than oil for delivery next week, next month or next year. If there is backwardation between two months of +$1/bbl, this means that the earlier month trades at a premium of $1/bbl to the later month. |
| B/D or b/d | Barrels per Day |
| B/L | See bill of lading |
| Barrel | A unit of volume used for petroleum and refined products. 1 barrel = 42 US. gallons. |
| BCTI | Baltic Clean Tanker Index |
| BDTI | Baltic Dirty Tanker Index |
| Bear Market | A market where participants anticipate that prices will go down, i.e. that the level of the forward oil price curve will decrease. |
| Benchmark | A grade of crude oil that is traded at a fixed and flat price and which provides a price reference point for formula priced transactions. See also Marker crudes. |
| Bill of Lading (B/L) | A document that records the details of a shipment such as the Consignor, the Consignee, the loading date, the amount loaded and brief quality specifications. It also can be used as a document to prove that the holder has a legitimate title to the cargo in question. The legal status of the B/L as a document of title is subject to debate. One legal viewpoint is that the B/L is no more than a receipt for a cargo. Irrespective of the legal debate, the oil industry treats B/Ls as if they were documents of title and they are therefore very valuable documents whose security should be guarded as closely as that of money. B/Ls are typically provided in three originals for historic reasons. If three B/Ls exist all three must be in the possession of the party claiming title to the oil. |
| BITR | The Baltic International Tanker Routes. Reports on a number of the most popular international routes each day at a given time. |
| BNOC | British National Oil Corporation |

| | |
|---|---|
| Book-out | Booking out involves accomplishing a contract with a cash payment rather than with physical delivery. This occurs when a "dry loop" is identified within a chain of contracts for one parcel of oil, which may be a tanker-loaded cargo or a pipeline transfer. For example, say Company A, is identified mid-chain as a seller of the cargo, but turns up again later in the chain as the buyer of the same cargo. With the cooperation of all the companies between Company A appearing as the seller and Company A appearing as the buyer, that part of the chain can be uncoupled and cash-settled. To achieve the book-out each party agrees to a reference volume and a reference price for the cargo or pipeline transfer. Each party then pays or receives payment from the other parties to its own purchase and sales contracts a sum of money amounting to the difference in price between the price applicable to its own contract and the reference price agreed by the book-out participants. This process minimises operations, because no nominations and no wet cargo or pipeline transfer have to be passed down the booked out portion of the chain. The booking out of dry loops minimizes the amount of money that has to be moved between parties, thereby reducing bank and LC costs: the cash transfer is the difference between the reference price and the sales price under the contract, not the whole value of the cargo or pipeline transfer. |
| Brent Blend Crude Oil | UK Brent Blend is a blend of crude oil from various fields in the East Shetland Basin. The crude is landed at the Sullom Voe terminal and is used as a benchmark for the pricing of much of the world's crude oil production. |
| Brent Market | A widely traded oil market in North Sea Brent Crude that emerged in the early 1980s. It is used as a key source of international price risk management and as a benchmark for crude oil pricing under both term and spot transactions. The Brent market is made up of physical, forwards, futures, swaps and options contracts. |
| Brent NX | Brent New Expiry. A futures contract introduced by ICE in 2012 to accommodate the move from 21-Day BFOE to 25-Day BFOE in the forward market. |
| BS&W | Bottom sediment and water, often found in crude oil and residual fuel. |
| Bull Market | A market where participants anticipate that prices will go up, i.e. that the level of the forward oil price curve will increase. |
| Call option | The holder (buyer) of the option has the right, but not the obligation, to buy the oil at a given (strike) price, by (or on) a certain date (expiry). |

| | |
|---|---|
| CAPEX | Capital Expenditure |
| Carbon | The base of all hydrocarbons; capable of combining with hydrogen in almost numberless hydrocarbon compounds. |
| Catalyst | A substance which aids or promotes a chemical reaction without forming part of the final product. It enables the reaction to take place faster or at a lower temperature, and remains unchanged at the end of the reaction. |
| Catalytic Cracking | The refining process of breaking down the larger, heavier and more complex hydrocarbon molecules into simpler and lighter molecules. Catalytic cracking is accomplished by the use of a catalytic agent and is an effective process for increasing the yield of gasoline from crude oil. |
| CEO | Chief Executive Officer |
| CFD Market | Contract for Difference Market. A similar market to the DFL market. It involves trading the fixed for floating swap price differential between Dated Brent and the 25-Day BFOE forward contract. |
| CFO | Chief Financial Officer |
| CFTC | US Commodities Futures Trading Commission |
| Clearinghouse | An organisation responsible for settling trade accounts for financial transactions including cleared swaps and futures. |
| CME | Chicago Mercantile Exchange |
| Condensate | A term used to describe light liquid hydrocarbons separated from crude oil after production and sold separately. |
| Consignee | The party identified on the B/L as the receiver of a cargo. |
| Consignor | The party identified on the B/L as the supplier of a cargo. |
| Contango | The forward oil price curve slopes up from left to right meaning that the price of oil for delivery tomorrow is less expensive than oil for delivery next week, next month or next year. If there is contango between two months of -$1/bbl, this means that the earlier month trades at a discount of $1/bbl to the later month. |
| Contract of Affreightment (COA) | A contract to charter a ship from a pool of qualifying vessels on demand. |
| Cost and Freight (CFR) | A term which specifies with whom shipping and loading costs lie and where the title for the goods passes from seller to buyer. The seller provides the ship and pays the cost of freight to transport the oil to the buyer's discharge port. |

| Cost Insurance and Freight (CIF) | A term which specifies with whom shipping and loading costs lie and where the title for the goods passes from seller to buyer. Typically the seller provides the ship and pays the cost of insurance and freight to transport the oil to the buyer's discharge port. |
|---|---|
| Cost Recovery | A contractual right for a production firm to hold back revenue from the sale of the oil to recover some of the expenditure used to develop the field. |
| Crude oil | Oil produced from a reservoir after associated gas has been removed by separation. A fossil fuel formed from plant and animal remains many millions of years ago, it comprises organic compounds built up from hydrogen and carbon atoms and is accordingly often referred to as a hydrocarbon. Crude oil also contains small quantities of oxygen, nitrogen and sulphur and other elements and compounds. |
| Dated Brent | A term for a physical cargo of Brent Blend crude which has received its loading date range from the Sullom Voe terminal operator. The Dated Brent market refers to the market in Brent Blend cargoes loading up to 25 days forward. |
| Deadfreight | Unused tanker capacity, which must be paid for by the charterer whether used or not. |
| Deadweight Tonne | Used as a broad measure of the size of a ship. It measures the displacement of the vessel fully laden with cargo, crew, fuel, ballast etc. |
| Delivered Ex Ship (DES) now replaced with Delivered at Port or Delivered at Place (DAP). | A term which specifies with whom shipping and loading costs lie and where the title for the goods passes from seller to buyer. Typically the seller charters the vessels and retains risk and title to the oil until it is delivered at the port of discharge. |
| Demurrage | The charges which must be paid by the culpable party if there is a delay in loading or discharge beyond the laytime allowed. |
| Desalting | Removal of salt from crude oil. Salt is corrosive therefore its presence, over specified limits, has an impact on the value of crude oil. |
| DFL Market | The Dated-to-Front-Line Market. A similar market to the CFD market. It involves trading the fixed for floating swap price differential between Dated Brent and Futures Brent. This market cash settles by reference to monthly, quarterly or annual price averages. |
| DFSA | Dubai Financial Services Authority |
| Distillation (Fractional distillation) | A separation process based on the difference in boiling points of the liquids in the mixture to be separated. Successive vaporisation and condensation of crude oil in a fractionating column will separate out the lighter products, leaving a residue of fuel oil or bitumen. |

| Dodd Frank Act | The US Dodd-Frank Wall Street Reform and Consumer Protection Act. |
|---|---|
| Downstream | Considering the chain of supply from the wellhead to the end consumer, the furthest point upstream is the crude oil production end of the chain and the furthest point downstream is the sale of refined products to end users. |
| Dubai Benchmark | A crude produced in Dubai, one of the United Arab Emirates. Dubai is commonly used as a reference price for the Asia-Pacific region. |
| Dubai Mercantile Exchange (DME) | A commodities exchange based in Dubai, offering a futures contract in Oman crude oil. |
| E&P | Exploration and Production. |
| EFP | Exchange for Physical. The transfer of a physical contract onto a regulated futures exchange. |
| EIA | Energy Information Administration. A statistical and analytical agency within the US. Department of Energy. |
| ESPO pipeline | Russian East Siberian Pacific Ocean Pipeline. |
| Feedstock | The blend of hydrocarbon input into a refinery. |
| Floating Production and Storage Operation (FPSO) | This is an operation that is undertaken by a group of joint venture partners for producing and storing hydrocarbons in a tanker that has been converted or built for the purpose. FPSOs provide economic production solutions in areas where there is a lack of transportation infrastructure (such as a pipeline) and the reserve estimates do not justify capital expenditure on providing such infrastructure. |
| Forward Contract | An OTC commitment between a seller and a buyer to deliver and take delivery of a cargo of a specified grade of oil during the course of a specified forward period, usually of a month, at a fixed price. |
| Forward Market | A physical market in cargoes for delivery in a future month. |
| Forward Oil Price Curve | A snapshot taken at a particular instant in time of the prices at which buyers and sellers are actually prepared to deal at that moment in time in oil for delivery at different dates in the future. |

| Four O'clocking (See also 25-Day BFOE) | A phenomenon of the 25-Day BFOE market. The seller of such a cargo is required to inform the buyer of which of the four grades of crude oil it will receive and on which three day loading date range the cargo will be delivered in the specified month by 4pm London time 25 clear days before the first day of the three day loading date range. If this deadline is not met the cargo ceases to qualify as a 25-Day BFOE cargo and is instantly transformed into a Dated cargo. Whoever is holding the cargo as the atomic clock strikes 4pm is said to have been Four O'clocked. This has negative financial consequences in a contango market, because the holder of the cargo will have to sell it at a lower "Dated" price and buy another cargo that does qualify for supply into the 25-Day BFOE market at the higher 25-Day BFOE price. |
|---|---|
| Free on Board (FOB) | A term which specifies with whom shipping and loading costs lie and where the title for the goods passes from seller to buyer. Typically the buyer charters the vessel and risk and title to the oil passes from the seller to the buyer at the load port. |
| FFA | Freight Forward Agreement. An OTC forward contract in ships. |
| FSA | The UK Financial Services Authority |
| G | The grade differential. The difference between the price of a particular grade of oil and that of its closest benchmark grade. |
| GDP | Gross Domestic Product |
| GHG | Greenhouse Gas |
| GPW | Gross Product Worth is the weighted average value of all refined product components (less an allowance for refinery fuel and loss) of a barrel of crude oil. GPW is computed by summing the the value of products calculated by multiplying the price of each product by its percentage share in the yield of the total barrel of crude. |
| GTC | Industry specific general terms and conditions of sale. |
| Hedging | A strategy to reduce the uncertainty of future revenue due to oil price volatility. A risk-reducing strategy. It is not necessarily a revenue maximisation strategy. |
| Hydrocarbons | Organic compounds consisting of hydrogen and carbon. They may exist as solids, liquids or gases. |
| ICE | The InterContinental Exchange, headquartered in Atlanta, Georgia, provides standardised Brent futures contracts and contracts in other oil products and in other commodities. |

| | |
|---|---|
| IEA | International Energy Administration. An international institution founded in 1973/4. It provides autonomous information, analysis, research and recommendations to the international community on energy matters. |
| IEF | International Energy Forum |
| Implied Volatility | The level of volatility required to generate an option premium given a known market price for the underlying oil, a known interest rate, a known expiration date and a known strike price. It is short-hand for all other factors that have an influence on the premium for the purchase or sale of an option with a given strike price. |
| INCO terms | International Chambers of Commerce Rules for the use of domestic and international trade. |
| Initial Margin | A good faith dealing deposit required by clearinghouses from users when a position is opened on an exchange. This is returned to the user when the position is closed and all obligations to the exchange have been discharged. |
| Integration | A term that describes the extent of any one given company's involvement in all the various stages of the oil industry from exploration and production, to pipelines, shipping, refining, marketing and distribution. |
| IOSCO | International Organisation of Securities Commissions. |
| IPE | International Petroleum Exchange. Taken over by ICE in 2001. |
| IRF | Inside the Ring Fence. See also ring fencing. |
| ISDA | The International Swaps and Derivatives Association |
| Joint Operating Agreement (JOA) | An agreement that regulates the exploration and development of a commonly held acreage or license concession by a number of firms acting together. It normally determines which company will be the main operator of the acreage on behalf of the group and outlines voting requirements, financial systems, operational procedures and decision-making within the group of joint venture parties. |
| Joint Venture (JV) | A group of companies operating together to explore for or produce petroleum. The terms of the partnership are defined in the Joint Operating Agreement. |
| Kinematic Viscosity | A measure of how easily a fluid flows. The technical definition is the absolute viscosity divided by the density. |

| Laycan | A commonly misused term. It is often used to mean laydays, but it actually means the time at which a charterparty can be cancelled if a vessel fails to arrive within its laydays. |
|---|---|
| Laydays | The agreed range of days within which a vessel should arrive to commence loading or discharge of a cargo. |
| Laytime | The time allowed in a charterparty for a cargo to load and discharge. It is typically 36 running hours but may vary with the size of the tanker and the cargo. |
| LC | Letter of Credit |
| Licence Agreement | A licence granted to an organisation to explore acreage for the presence and recovery of hydrocarbons. The licensor is usually a government or NOC and the licensee may be a company, or group of companies, with technical exploration expertise and financial capability. |
| Lifting Agreement | A contract defining the terms under which fields may use common storage facilities for the aggregation of daily production from one or more fields into transportable cargo lots. It determines the rights of producing companies to take delivery of cargoes of oil produced in a fair and equitable order within the constraints of the facilities. |
| LP | Linear programme. Employed by refineries to assess the maximisation of the refining margin from processing a range of different grades of crude oil. |
| LOI | Letter of Indemnity |
| Long Term Charter (time charter) | A contract to supply a vessel to a charterer for a period of time. |
| LOOP | The Louisiana Offshore Oil Port. A transhipment point in the US Gulf Coast where large cargoes can break bulk for supply into the US domestic market. |
| Major oil company | A term broadly applied to those multinational oil companies, which by virtue of size, age, or degree of vertical integration are among the pre-eminent companies in the international petroleum industry. |
| Marker Crudes | Benchmark grades of crude oil against which other crudes are priced. Widely used marker crudes include West Texas Intermediate, Brent Blend and Dubai. |
| MiFID | The European Market in Financial Instruments Directive. |
| MOLOO | More Or Less at (ship) Owner's Option |
| Mt | Metric tonnes |
| Naphthenes | Also known as cyclo-paraffins or C-alkanes. Saturated cyclical chains of more than 4 carbon atoms. |

| | |
|---|---|
| Netback Price | Crude oil priced by reference to the market value of the refined products that can be extracted from it in the refining process. |
| NOC | National Oil Company |
| NOR | Notice of Readiness. A formal tendering of notice by the master of a vessel that it is ready to come into the port to commence loading or discharge. It is a key notification in the calculation of laytime and demurrage. |
| NPV | Net Present Value |
| NYMEX | The New York Mercantile Exchange. A platform trading crude oil and oil products. Owned by the Chicago Mercantile Exchange since 2008. |
| OECD | The Organisation for Economic Cooperation and Development. Member countries are Australia, Austria, Belgium, Canada, Chile, Czech Republic, Denmark, Estonia, Finland, France, Germany, Greece, Hungary, Iceland, Ireland, Israel, Italy, Japan, South Korea, Luxembourg, Mexico, Netherlands, New Zealand, Norway, Poland, Portugal, Slovak Republic, Slovenia, Spain, Sweden, Switzerland, Turkey, United Kingdom and the United States. |
| Oil Terminal | Facilities used for stabilising and storing oil and for loading and unloading tankers. |
| Olefins | Also known as alkenes. Unsaturated, unstable compounds that are made up of hydrocarbons containing carbon double bonds. Do not occur naturally within crude oil but are released in refineries by the cracking process. |
| OPEC | Organisation of the Petroleum Exporting Countries. Members are the Republic of Iran, Iraq, Kuwait, Saudi Arabia and Venezuela, (1960), Qatar (1961), Libya (1962), the United Arab Emirates (1967), Algeria (1969), Nigeria (1971), Ecuador (1973), and Angola (2007). Indonesia was a member, but it left in January 2009 after it became a net importer. Gabon was a member until 1995. |
| OPEX | Operating Expenditure |
| ORF | Outside the Ring Fence. See also ring fencing. |
| OSP | Official Selling Price. Usually refers to the price announced or negotiated by a NOC for use as a tax reference price or for cost recovery and profit sharing in PSCs. |
| OTC Market | Over-the-Counter transactions occur when two parties negotiate the sale of a cargo of crude oil directly without using one the standardised products that are traded on regulated exchanges, such as ICE or NYMEX. This involves bipartite contracts with two identified counterparties. Each is exposed to the other in the contract for performance and payment. |

| Paraffins | Also known as alkanes. Straight (normal) or branch (iso-) chained hydrocarbons, which are "saturated" with hydrogen and which are therefore stable compounds. |
|---|---|
| Petroleum | A term applied to crude oil and oil products in all forms. |
| PONA Number | The components of crude oil. Stands for the relative content of Paraffins, Olefins, Naphthenes and Aromatics within the crude oil. |
| Posted Price | An announced or advertised price indicating what a firm will pay for a commodity or the price at which the firm will sell it. |
| ppm | Parts per Million |
| PRA | Price Reporting Agency |
| Premium (option) | The non-refundable amount of money paid upfront by an option buyer to an option seller. |
| Premium (price) | A positive price differential |
| Production Sharing Agreement or Contract | A Production Sharing Contract (PSC) is a contract that lays out the conditions for the extraction of hydrocarbons from a licence area. The terms cover tax, royalty, cost recovery and revenue distribution from the production of oil and/or gas. It is typically negotiated between a government and a group of production companies although it can be a bilateral agreement between a national government and a single production entity. |
| Prompt Barrel | Physical crude oil for immediate delivery. |
| Put option | The holder (buyer) of the option has the right, but not the obligation, to sell the oil at a given (strike) price, by (or on) a certain date (expiry). |
| PVT | Pressure Volume and Temperature Analysis |
| Quality Bank | Similar to a VAM. A mechanism for attributing fair value when two or more grades of crude oil are commingled in a pipeline or in storage. The contributor of the superior quality of oil is compensated by receiving a cash payment from the contributor of the grade of lesser quality. |
| Ring Fencing | The isolation of costs and revenue from one field or licence area from the rest of a company's assets or trade. Used in PSCs to ensure that only those costs that are attributable to the field under consideration are recovered from the revenue stream of that field. |
| Slate | The range of grades of crude oil processed by a refinery. |
| SOMO | The state oil marketing organisation of Iraq. |
| Sour Crude | Crude oil with a high sulphur content. |
| SPAN | Standard Portfolio Analysis of Risk software. This is used to determine the riskiness of a contract and to set the initial margin that must be paid by the trader. |

| Specific Gravity | The ratio of the density of a material to the density of water. |
|---|---|
| Strike Price | The price at which the buyer of an option is entitled to sell (put) or buy (call) the underlying commodity. |
| SUKO 90 | General Terms and Conditions used in the forward contract for Brent. First issued in 1981 by Shell UK. The 1990 update is still in use in the 25-Day BFOE market. |
| Sweet Crude | Crude oil with a low sulphur content. |
| T | The time differential. Represented by the slope of the forward oil price curve. A curve that slopes up from left to right is in contango and T has a negative value. A curve that slopes down from left to right is in backwardation and T has a positive value. By convention T is calculated by deducting the price of oil for future delivery from the price of oil for prompt delivery. |
| Tapis | A grade of crude oil used as a benchmark in Asia. Produced in Malaysia, Tapis is a very light (46°API) and very low sulphur crude oil (<0.03%), typical of the region. |
| TBP | True Boiling Point. A curve showing the amount of a particular grade of crude oil that boils off at increasing temperatures. |
| Time value | A valuation of how long the buyer of the option is given to decide whether or not to exercise the option before it expires. |
| Total Acid Number (TAN) | The quantity in milligrams of potassium hydroxide that is required to neutralise all acidic constituents present in a 1 gram sample of crude. |
| Ullage | Free capacity for the storing and transportation of crude oil. |
| Upstream | Considering the chain of supply from the wellhead to the end consumer, the furthest point upstream is the crude oil production end of the chain and the furthest point downstream is the sale of refined products to end users. |
| VAM | Value Adjustment Mechanism. A mechanism for attributing fair value when two or more grades of crude oil are commingled in a pipeline or in storage. The contributor of the superior quality of oil is compensated by receiving additional barrels of the blended oil from the contributor of the grade of lesser quality. |
| Variation Margin | The amount of money that is paid to or received from an exchange's clearinghouse daily, reflecting the change in the profitability of an open position each day. It is based on a "mark to market" price calculation. |

| | |
|---|---|
| Volcker Rule | A section of the Dodd Frank Act, which has been interpreted as preventing US banks from proprietary trading with clients' money or from advising clients while at the same time trading their own book. A complex debate is on-going concerning the actual intention and implementation of this rule. |
| Wet Barrel | A physical barrel of crude oil or refined product. This compares with a "paper barrel" for which physical delivery is not anticipated and it is expected that the paper contract will be settled in cash by entering into an equal and opposite position before the delivery date. |
| Worldscale | "New Worldwide Tanker Nominal Freight Scale," An international freight index for tankers that provides a method of calculating the cost of the transporting oil by a reference ship at a reference speed under reference sea conditions for a single voyage between two ports. |
| WTI | West Texas Intermediate is a grade of sweet, light crude oil. It is one of the physical grades of oil deliverable into the futures contract of the New York Mercantile Exchange at Cushing, Oklahoma. |
| Yield | The proportion of heavy or light products which can be derived from a given barrel of crude oil. |
| ZCC | Zero Cost Collar. The financing of the purchase of one option by selling another option. The strike price of the options bought and sold are set so that the premium paid for buying one is exactly offset by the premium received from selling the other. |